Paolo Nanetti

Coatings from A to Z

A concise compilation of technical terms
2nd revised edition

Paolo Nanetti: Coatings from A to Z

Cover: puckillustrations/Fotolia.com

Bibliographische Information der Deutschen Bibliothek
Die Deutsche Bibliothek verzeichnet diese Publikation in der Deutschen Nationalbibliographie; detaillierte bibliographische Daten sind im Internet über http://dnb.ddb.de abrufbar.

Paolo Nanetti
Coatings from A to Z
Hanover: Vincentz Network, 2016
European Coatings Library
ISBN 978-3-86630-638-7

Please ask for our book catalogue
Vincentz Network, Plathnerstr. 4c, 30175 Hanover, Germany
T +49 511 9910-033, F +49 511 9910-029
books@european-coatings.com, www.european-coatings.com

Layout: Vincentz Network, Hanover, Germany

ISBN 978-3-86630-638-7

European Coatings Library

Paolo Nanetti

Coatings from A to Z

A concise compilation of technical terms
2nd revised edition

Contents

Foreword

This compact technical lexicon is meant as an easy to use aid, made available to all interested people in need of a quick and concise source of information about single headwords of the coating technology.

It neither presupposes special coating knowledge nor fundamental chemistry cognitions. Nevertheless, as a compact reference book it is suitable to recall latent coating knowledge, for example what has been acquired in seminars and informative conferences or other educational meetings.

Its use is also recommended where knowledge gaps have to be filled quickly and selectively, for example at customer meetings, trade events, and professional presentations.

For readers with a technical education in coatings, this book can also be helpful as a compact and easy to carry information source, when it may be necessary to have a quick overview of the headwords from different sub-fields of the coating technology and to link single terms so that they contribute to a fuller understanding of the field.

In order to improve the general comprehension of this lexicon, chemical formulas and complex scientific explanations have been deliberately avoided. To further increase its practical usefulness for the reader, the lexicon is enriched with many examples of technical applications and practical implications, as well as helped by the creation of cross references to similar terms.

Finally, attached to this book is a register of all German terms with their corresponding English terms and vice versa, in order to facilitate its use as a means for technical translations from one language to the other. The most efficient use is if the reader also has available the German version "Lack von A bis Z".

I would like to express my deepest gratitude to Professor Thomas Brock of the University of Applied Sciences Niederrhein at Krefeld/

Germany for the technical assistance he has lent in the review of the text and for his suggestions concerning the compilation of the text headings. Finally, a special thank goes to my sister Raffaella Nanetti-Leonardi, who has been in the United States for over thirty years and is a professor of urban planning at the University of Illinois at Chicago. Her help has been critical for the correction and optimisation of the English text.

I wish the readers much satisfaction in their use of this book, together with a bit of fun as they nurture their curiosity with information on the fascinating world of coatings and paints.

Paolo Nanetti

Zülpich/Germany, February 2016

A

abrasion resistance: Important technological property of coatings, which are subject to friction wear. This concerns in the first place floor paints, road line paints, under-sealing coatings, but also wood enamels and interior paints.

accelerators: Additives, which are able to accelerate definite chemical crosslinking reactions and therefore to shorten the hardening times. As examples may be mentioned: driers for oxidative drying (→ *driers*) and cobalt octoate for unsaturated polyesters.

acceptance test: Check of raw materials, delivered from suppliers, before their employment in the coating production. The goal of acceptance tests is to warrant the compliance with defined technical values. To control costs, the emphasis previously placed on acceptance tests has been greatly reduced. Nowadays coating companies rely more and more on the test reports of the raw material suppliers.

achromatic colour: → *colour, achromatic*

acid curing: It is the principle of an acid-catalysed condensation reaction, which can take place with all types of resins containing methylol-groups, formed by a reaction with formaldehyde (→ *amino resins* and *phenolic resins*). The advantage of the acid-catalysis is that the reaction takes place at room temperature, thus without stoving. This is important for bulky machine-casings and other industrial devices, and for parquet floors wood boards as well. Theoretically, all of the so called mineral acids (sulphuric, hydrochloric, nitric acid etc.) may be employed as catalysts. However, if iron corrosion is expected, the passivating phosphoric acid or the mild paratoluene-sulphonic acid are mostly used.

acid curing coating: → *acid curing*

acid resistance: Property required of coating systems as well as of their components (for example pigments), if they are exposed to substances which react like acids, when they are used. This property is particularly important in the case of highly chemical resistant coating materials, expressly produced for employment in the chemical sector, or in the interior can-coating (→ *gold coatings*) sector, but also in coatings which are exposed to polluted atmosphere (acid rain) during weathering.

acids: Strictly speaking they are chemical compounds, which in a solution, due to a protolysis reaction with the medium, separate positively charged hydrogen ions (protons). They represent the counterpart of the → *bases* and can be neutralised by the latter. In the coating technology the acids play a multiple role, because on the one hand they are important source materials for the production of binders (phthalic acid, maleic acid, different fatty acids, acrylic and methacrylic acid), while on the other hand they are employed as catalysts for the → *acid curing*. Further important properties of the acids are their function as passivating and reactive components in one- and two-component primers (→ *primers*) and their use as neutralisation agents in cathodic electrodeposition coatings (→ *electrodeposition coating, cathodic*).

acid value: Important chemical value, which is used above all in the polyester synthesis or in the quality control of polyesters (→ *polyester resins*). It expresses how many milligram of potassium hydroxide are needed, in order to neutralise one gram of resin. The evaluation takes place by means of a titration, with phenolphthalein as indicator. The acid value indirectly gives evidence of the free acid groups which are already present in the resin (→ *carboxyl group*) and, therefore, also of the structure and the reactivity of the resin.

acrylic resins: Versatile group of binding agents with highest performance relative to light and weather resistance. Above all, acrylic resins play an indispensable part when only the smallest possible colour changes can be accepted, and also after very long periods of weathering and influence of ultraviolet rays. Acrylic resins are produced by polymerisation of derivatives of acrylic or methacrylic acid (→ *monomers*). Depending on the chemical structure of the involved monomers, acrylic resins can be employed for physical drying, for cross-linking with isocyanates in two-pack systems or for baking enamels, and additionally for ultraviolet curing lacquers. The main fields of application are automotive coatings and vehicle refinishing, aircraft and commercial vehicle coatings, high performance house paints and synthetic material plasters.

addition polymerisation: Chemical reaction which underlies the curing of two-pack coating systems (for example, isocyanate-based enamels). During the reaction a stepwise cross-linking of both reactants takes place without separation of secondary products. By the progression of the chemical crosslinking, with the exception of two-pack water-borne coatings, the solidification of the coating material also increases (→ *pot life*). After termination of the reaction the hardened coating film is very resistant and not soluble any more.

additive: Auxiliary substance, component of coating materials, which is added to the main ingredients of paint formulations only in little quantities, mostly between 0.1 % and 3 %. The task of the additive is to influence, in a very specific and selective way, some special coating properties. As additives have a very high efficiency, they must be dosed carefully to prevent negative appearances. To make their exact dosing easier, additives are used in the form of highly diluted solutions or pastes.

additive colour mixture: Method to generate colours by mixing coloured lights. What the human eye perceives as a white light, is in fact a mixture of coloured spectral lights, whose colour depends on the respective wavelength of the radiation. One can decompose a standardised white light in three likewise standardised basic colours: violet-blue, green and orange-red. If one mixes all three basic colours, in the form of lights with the same intensity, a white light will be generated. If one mixes only two of them, a mixed light will be obtained, whose colour will be brighter than that of the single original components; for example from green and orange-red one will get a brighter yellow, or from violet-blue and green a brighter blue-green. The importance of the additive colour mixture for the coating sector is limited, being practically restricted to the functioning of the pearlescent pigments. On the contrary, the mixture plays a very important part in the lighting and television technique.

adhesion: Interaction of intermolecular forces of attraction, mostly physical ones, between different substances. These forces are responsible, among other things, for the adhesive strength of adhesives and coatings on substrates.

adhesion primer: → *primer*

adhesion tests: The most commonly used measuring method, to determine the adhesive strength in the coating technology, is by far the → *cross-hatch adhesion test*, because it can be carried out without complexity and it provides significant results. The pull-off test is used very infrequently. In regard to the pull-off test, the adhesive strength is measured by means of a disc bonded to the surface and of a tensile testing machine.

adhesive strength: Prime property of all coating materials and systems. Without sufficient adhesive strength, in the long run the other required coating properties cannot be performed. The adhesive strength is based on different physical attractive forces of electrical nature like Van der Waals forces, dipole forces and hydrogen bridge bonds (→ *adhesion*), which act between two substances, for example primer and substrate or topcoat and filler, subject to their polarity. Basic requirements for an optimal adhesive strength are the accurate pre-treatment (degreasing, removing of silicone remains, sanding etc.) as well as the sufficient wetting of the substrate.

aeration: → *ventilation*

aerosol: Fine dispersed liquid droplets in a gas, generally air. During the spray application of liquid coating materials, aerosols are artificially generated by atomisation, which provides an uniform distribution of the coating droplets all over the object surface. Coating aerosols must not be inhaled by the users, because they can cause health damage.

aerosol lacquers: Coatings, which are formulated for application by means of spray cans. The total compatibility of the coating material with the liquid propellant agent contained in the can (in the past fluorinated/chlorinated hydrocarbons, but today in Europe and America only propane/butane mixtures or dimethyl ether), is of major importance. Another important requirement is that the pigments should not form hard cake sediments even after long storage periods, in spite of the low viscosity in the aerosol can. They have to be stirred up completely by simple shaking of the can and by means of the metal balls contained in the can, to permit a perfect retrieving of the desired colour. While aerosol lacquers dry very quickly, nevertheless they have to show good levelling. If they are destined to application in automotive refinish, they have to be suitable for polishing, even after a short drying. The most important film formers for such coating systems are nitrocellulose and physically drying acrylic resins.

after-treatment: In the coating technology, this term refers essentially to processes aimed at improving some properties of pigments and fillers (extenders). Thus, for example, all titanium dioxides for the coating industry are after-treated to provide a reduction of their tendency to chalking. Difficult to disperse channel carbon blacks are after-treated by oxidation of their surface, in order to get better wetting properties. An after-treatment variant, which can be found often in the extender-sector, is the coating with silanes. In this case, the particle surface is equipped with water repellent groups, in order to improve the → *barrier-effect* in corrosion protection systems.

agglomerates: Coarse particles of pigment, which develop by a disorderly association of primary particles and → *aggregates*. The agglomeration takes mainly place by the drying of the pigment, during its production process, but it occurs also as flocculation, during the storage and the processing of coatings. This depends on the effects of the forces of attraction, situated on the surface of the little particles involved, which stick to each other along edges and corners. Because of the agglomeration the tinting strength of the pigment worsens, as well as its colour purity. The aim of the dispersion process in the coating production is to divide as efficiently as possible the agglomerates into primary particles and little aggregates, in order to achieve the full performance of the pigment (→ *dispersion*).

aggregates: Pigment particles, which consist of associated primary particles aligned side by side. Since their optical properties are very similar to those of primary particles (with the exception of a somewhat lower tinting strength), aggregates disturb the process only when they are so big as to influence the gloss of the coating material in a negative way. As a rule they are not destroyed during the dispersion process (→ *dispersion*).

agitator: → *dissolver*

airbrush technique: Special method of application, suitable for decoration and design uses. Therefore, little spray guns are employed, which differ only a little from high pressure spray guns (→ *high pressure spraying*). However the nozzles of the airbrush spray guns produce merely point shaped spray jets, which make possible to "draw" with them. The diameters of the airbrush nozzles are between 0.2 and 1 mm and the air pressure between 1 and 1.5 bar.

airless spraying: It is also named "hydraulic spraying". In this application process no compressed air is employed. The atomisation is caused by forcing the coating material at a high pressure (200 to 600 bar) through a fine nozzle (0.5 to max. 1mm). When leaving the nozzle the coating material is subject to an abrupt drop in pressure and to extreme turbulences, which tear it apart and atomise it immediately. However, the atomisation is not as good as spraying with compressed air (→ *high pressure spraying*), because by airless application the coating droplets remain coarser and the spray jet shows a sharp boundary. This causes an erratic surface appearance. On the other hand there are many advantages to this process, such as less overspray, high application efficiency and the possibility to apply thick coating films with each spraying operation.

airmix spraying: Airless atomisation with air assistance. In this process compressed air of 0.5 to 2 bar is used, in order to improve the relatively bad airless atomisation. On the other hand the coating pressure has to be reduced to 20 to 80 bar, otherwise a too high over-spray would arise.

The surface appearance is a compromise result, between airless and high pressure application.

air separation: Method for the separation of powdery substances into fractions, according to their particle sizes. The powder is put into an air stream which leads, by means of pipelines, to separation devices (cyclones) situated at different distances. Since with the decrease in the solid particle diameter the influence of the air resistance on the mass increases, coarser particles sink to the bottom earlier than finer ones. Thus a separation into different particle fractions is made possible, which are gathered in different cyclones. Air separation is above all employed for the preparation of fillers (extenders), which are produced from ground natural minerals.

alcohols: After water they are the most polar solvents. However, the polarity decreases with the increasing size of their molecules. While the "lighter" alcohols like ethanol or isopropanol are high polar, and thus completely water miscible, the "heavier" butanol is only up to 4 % water miscible. According to their polarity alcohols play an important part in water-borne coatings or as solvents for polar binders, like nitrocellulose, polyvinyl butyral, urea and melamine resins.

aliphatic hydrocarbons: Group of solvents, consisting of chains of carbon and hydrogen atoms. They are also named spirits (specialist spirits, white spirits) and represent the solvents for the coating industry with the lowest → *polarity*. Due to this property they are only able to dissolve non-polar film formers like fatty oils, fatty acids, long oil alkyd resins etc.. As they have a mild odour and are less harmful for the human body they are the first choice for painting in interior decorating and for use in architectural materials.

alkali resistance: Resistance against bases. This property is particularly important for coatings which are in frequent contacts with substances producing an alkaline reaction (→ *saponification*), like detergents and burnishes.

alkyd resins: Collective name for all oil modified polyesters. They represent one of the most important binder classes. The structure of the alkyd resins is based upon the idea to link up through a chemical process the elastic, chain-shaped molecules of the fatty acids, which are contained in the natural fatty oils, with the rigid, hard backbone of a polyester. Thus an internal plastificisation of the resin takes place. If the fatty acids of the employed oil are sufficiently unsaturated, so they possess enough double bonds (→ *fatty acids*), an additional so called oxidative drying takes place, that is a chemical cross-linking by means of air oxygen. The drying of decorative and house paints is based upon this mechanism.

alkyd resins, long oil: They contain over 60 % oil. A sufficient hardening of films based upon such alkyd resins can only be achieved by an oxidative cross-linking of the fatty acid chains. Therefore, long oil alkyds are always modified with drying oils like linseed oil, wood oil, soya bean oil etc.. Since the oxidative drying develops relatively slowly, long oil alkyds are suitable for decorating type of applications with brushes or rolls. That is why they are mostly used in decorator's and house paints.

alkyd resins, medium oil: They have an oil content between the long oil and the short oil alkyds, that is between 60 % and 40 %. In terms of their application properties they build a bridge between the two just mentioned film formers. Depending on their oil content, medium oil alkyd resins are suitable for oxidative drying, in the higher oil range, or for stoving cross-linking, in the lower oil range.

alkyd resins, short oil: They contain less then 40 % oil. Short oil alkyds are almost exclusively modified with non-drying oils and are predominantly used for stoving systems in combination with amino resins or in nitrocellulose combination coatings.

alternating atmosphere test: It is a combination of the salt fog and saturated atmosphere test according to VDA 621-415, often also simply called VDA-test, which is above all prescribed in automotive specifications to test car and automotive accessories coatings in a practical way. There are three different variants, KK (constant climate), KFW and KTW, depending on whether the cycles in the climatic chamber, alternating with the salt fog test, take place with or without interruptions and with open or closed chamber door.

aluminium: For the coating technology aluminium has a twofold importance. On the one hand it is increasingly used as a material, because of its low density of only 2.7 g/cm^3 (almost 1/3 of the steel density) and of its weather resistance, thus representing an important substrate for coating processes. On the other hand it is the basic raw material for the production of the aluminium bronzes, which are the most important effect pigments for metallic coatings. As substrates for coatings aluminium and its alloys are relatively problematic in relation to the adhesion properties. A sufficient adhesion can be normally reached only by using an adequate pre-treatment, like chromating, or the employment of wash primers.

aluminium bronzes: Platelet-shaped effect pigments, based upon aluminium, for the employment in metallic coatings. Because of their production process, but also as a safety measure, aluminium bronzes are used in the form of pastes, which, besides the aluminium platelets, contain also grinding agents and solvents. Their manufacturing is carried out by forming into platelet-shaped particles a fine and very pure aluminium grit in steel ball mills. The grinding agents act in the same way as lubricants and prevent the Al-platelets from undergoing cold fusion. Depending of the type of grinding agent, bronzes with different floating properties can be obtained (→ *leafing* and *non-leafing bronzes*).

amines: Organic compounds with alkaline reaction (→ *organic chemistry*), which are formal derivatives of ammonia and act similarly to it. However, amines evaporate more slowly then ammonia and this represents a significant advantage for most applications in coatings. In water-borne systems they accomplish the important function of turning temporary suitable binders into either water-soluble or water-thinnable ones. A further important employment is the use of polyamines as hardeners in two-pack epoxyd systems.

amino resins: In this category fall all types of urea, melamine and benzoguanamine resins. They are derived from basic raw materials, which contain so called amino groups. By addition of formaldehyde to such amino groups, new reactive groups are generated (methylol groups), which can bring about a cross-linking by means of a condensation reaction, either among each others or with groups of external film formers. The reaction can take place either by stoving at higher temperatures or under the influence of acids at room temperature (acid curing coatings). Amino resins are the most important and many-sided partners for the formulation of stoving coatings.

anatase: One of two possible modifications of titanium dioxide with pronounced white pigment character. The crystal structure of anatase is built on in such a way that this white pigment develops by weathering a strong photo-catalytic effect on the binder molecules, thus contributing to the accelerated degradation of the film former matrix (→ *chalking*). Consequently, anatase is scarcely used in the field of paint. It plays a bigger part in the wall paper industry and in the pigmentation of special joint compounds and putties.

angle of incidence: Angle which a ray of light forms to the perpendicular, when it meets the interface between two different substances. The ray portion, which is reflected by the interface, represents the angle of reflectance and is identical to the angle of incidence (mirror reflection).

anion: Particle which bears electrical negative charges. It can consist of a single atom or of one group of different atoms. In the modern paint technology, the particle anion is above all important for electro-deposition coating processes (→ *electro-deposition, anodic*).

anode: Positively charged electrode, to which the negatively charged anions migrate to discharge in an electrolysis bath where direct current is applied. The deposition of electro-dipping systems is based on this electrolysis principle.

anti-corrosive pigments: They provide, through chemical or electrochemical reactions, the protection of metal substrates against corrosion. The chemical protection is mostly produced by the neutralisation of acid corrosion agents and/or by the forming of watertight metal soaps in the coating material. The electrochemical protection leads to different kinds of passivation layers, which have a very good adhesion on the metal surface and form a sort of protective skin on it (for example, by employment of zinc dust, zinc chromate and zinc phosphate). The prerequisite for an effective corrosion protection is a certain water solubility of the anti-corrosive pigments, because this is the only way to create the ions which are required for the protection mechanisms described above.

anti-floating agents: Contrary to their name, anti-floating agents are not capable of preventing the floating of pigments in coating. But they can help to hide the visible negative effects of floating, such as non uniform colour, formation of Bénard cells etc. (→ *floating*). These agents are surface-active substances like silicone oils (modified polysiloxanes), which, due to their very low surface tension, in the upper coating layers are capable of transforming the vertical streams, which are responsible for pigment separation, into horizontal ones. These streams flow into each other and intermix to produce an uniform colour.

anti-settling agents: Group of → *rheological additives*, which provoke an increase of viscosity in a resting coating material. As a result, the formation of cake hard deposits, caused by sedimentation of pigments and extenders, can be prevented or at least retarded. Important anti-settling agents are pyrogenic silicic acids, phyllosilicates (bentonites), cellulose ethers and derivatives of hydrated castor oil.

anti-skinning agents: Additives for oxidative drying, which are added to the coatings to prevent skinning during storage. The most commonly used substances are aldoximes, ketoximes or substituted phenols. These concentrate on the coating surface to perform their preventive function in a selective way, that is they form inactive chemical complexes with the metal atoms of the driers, which normally accelerate the cross linking reaction. The driers lose temporary their catalytic effects. After the coating has been applied, the anti-skinning agents evaporate and set the metal atoms free. Thus, the cross linking of the film can take place without delay.

anti-static agents: Substances which are able to increase the electric conductivity of coatings. Anti-static agents are added to coating materials, if these exhibit too low a conductivity level for electrostatic application.

anti-static cloth: → *sanding dust*

application methods: Methods of coating processing, which lead to films with specific technological properties. Generally a distinction is drawn between manual and industrial application processes. Important examples of manual application processes are: brushing, roll, → *high pressure spraying*. Examples of industrial applications are: automotive coating, electro-deposition and coil coating.

application technological properties: Properties of coating materials, specifically coating systems, which are connected with their processing. Primarily they are application and drying properties such as levelling, gloss, aptitude to brush application, hardness, scratch resistance as well as formation of effects by metallics; but they are also further processing and storage properties such as sanding, interlayer adhesion, sedimentation and frost stability. The coating factories have application departments, which are responsible for the testing of new products as well as for the processing of complaints.

application viscosity: It is the viscosity, which the coating material must have before starting application. It is normally adjusted shortly before processing, according to the directive of the coating manufacturer, mostly with original thinner supplied in addition to the coating material. Only in few instances, for example in the case of decorative paints for the brush application or some industrial paints, the viscosity of delivery corresponds already with the application viscosity.

aqueous dispersion: → *dispersion*

architectural paints: → *house paints*

areometer: Simple device for determining density of transparent liquids like solvents and clear coat materials. The areometer consists of a closed, hollow object of glass cylindrical in shape, which at its top ends into a narrow glass column. A graduation with the density values is attached to the glass column. The device is immersed into the liquid, which has to be tested. According to the density of the liquid, the glass column with the graduation emerges out of the liquid surface to a greater or lesser extent. The corresponding density value can be read directly at the phase boundaries between liquid and air.

aromatic-free: Demand which is increasingly lodged by customers of the coating industry and by the legislators, concerning the development of new products or the adaptation of old products to the state-of-the-art. The physiological harmfulness of the aromatic hydrocarbons, which in the past years provoked a drastic mark down of the →*MWC-values*, increasingly imposes the substitution of such important solvents like xylene for instance with butyl acetate or glycol-ether-esters.

aromatic hydrocarbons: Class of organic compounds, which are characterised by ring-shaped molecules and specific chemical properties. They always exhibit a benzene-nucleus in their structures. The aromatic hydrocarbons represent very important solvents for the coating industry. First of all toluene and xylene, but also higher boiling aromatic mixtures are of key importance. However, in the last few years because of new results from occupational medicine research, they have been meeting with mounting criticism as harmful coating components and thus they are increasingly substituted by less harmful solvents.

atomisation: Process, which leads to the decomposition of liquid coating materials into many little droplets, during their spray application. The higher the atomisation degree, the smaller the diameter of the so produced droplets. At a high atomisation degree, the optical quality of the coated surfaces usually improves. However, in some cases of too high an atomisation degree, in the painter's vocabulary also referred to as "dry spraying", it may cause worsening of the spray dust absorption and, consequently, damage to the coated surface, in addition to the increase of the spray dust loss (overspray). On the other hand, insufficient atomisation may provoke a non-uniform coated surface and, in an extreme case and depending on particular spray conditions, also real surface defects like orange peel effect or → *curtaining*. Regarding the different technical possibilities of application → *spray application*.

attrition mills: → *bead mills*

automotive coating: This term covers the industrial car coating at the assembly line (OEM-coating) as well as the → *automotive refinishing process*. The OEM-coating (OEM= Original Equipment Manufacturing) comprises different and mainly automated process steps, which are successively carried out in a finishing line. The process starts with degreasing and rinsing of the shell. This step is followed by activation and phosphatising with subsequent neutralisation and rinsing (→ *phosphatising*). The next step is the application of the anti-corrosive layer by

→ *electro-deposition coating*, followed by the application of the underbody sealing and of the filler, for the protection against stone chipping, with subsequent stoving. After eventual sanding of defective spots, the top-coating is carried out with electrostatic rapid-rotation atomisers. The stoving of the topcoat takes place at the end.

automotive (OEM*) fillers: Coating materials which, during the whole process of the car coating at the assembly line, have the function of protection against stone chipping and optimisation of the car body surface before the application of the top coat (→ *primer surfacers*). Today water-borne, so called “hydro-fillers” are also increasingly used. The hardening is carried out by stoving.

*OEM = Original Equipment Manufacturing.

automotive (OEM*) primers: Corrosion protection which is provided by the car coating at the assembly line. In the modern automotive coating technology, this aim is fulfilled exclusively by electro-deposition primers, which are applied by the ED-process (→ *electro-deposition coatings*). After the application the primers are stoved.

*OEM = Original Equipment Manufacturing.

automotive (OEM*) topcoats: The most commonly used variants are one-layer and two-layer topcoats. The one-layer coatings combine in a film all functions of the topcoat (colour, gloss, weather resistance etc.). In the two-layer systems, a division of tasks takes place. The pigmented under layer (base coat) is responsible for the colour or by metallics and pearls, for the effect formation The upper clear coat layer, which is then applied, protects against weathering and mechanical effects. Base coats are exclusively physically drying systems, while clear coats and one-layer coatings are stoving systems, based on acrylic film formers.

*OEM = Original Equipment Manufacturing.

automotive refinishing coatings: High-class coating materials to remedy coating damages or to renew automotive coatings, for primary coatings of buses and commercial vehicles, or for application of special finishes. They have to meet the highest claims about accuracy of shade, gloss, levelling and weather resistance. Unlike assembly line coatings (OEM), which are generally stoved, refinishing coatings are mostly two-pack systems, which are cross linked at room temperature or force-dried at a maximum of 80 °C. The reason for this difference in the coating process is that completely assembled cars, with full tank, tyres and electrical devices, could not withstand the stoving temperatures.

automotive refinishing process: It is exclusively carried out by manual methods. With the exception of the putty, which is applied with a spatula, the application of refinishing coatings takes place by means of different pneumatic spraying techniques (→ *high pressure spraying* and *HVLP*). Occasionally special methods, like → *air brush technique* for design and aerosol cans for do-it-yourself applications, are employed.

automotive refinishing systems: They consist of different coating materials, coordinated with each others, which in response to customer preferences can be used as a multi-layer system. The way they are combined with each others depends on the damage, which has to be remedied, that is the kind of the required repair or new coating. Important components of an automotive refinishing system are: putty (levelling of big substrate irregularities), wash primer (adhesion to the metal substrate), filler (levelling of small substrate irregularities, resistance to stone chipping), primer or primer-surfacer (corrosion protection), and top coat (gloss, weather and scratch resistance).

auxiliary solvents: → *co-solvents*

auxiliary products: → *additives*

azo-pigments: They number among the oldest, synthetic, organic pigments and are characterised by the presence of azo-groups (-N=N-) in their molecules. They show a very large hue range, which extends from yellow, through orange to red. Because they have a relatively low price, in comparison with other organic pigments, they are employed quite often. Disadvantages, though dependent on the type, include their rather mediocre light fastness and their mostly slow solvent resistance (→ *bleeding*).

B

ball mill: In the past it was one of the most important dispersing units in the coating technology. It consists of a lying cylindrical vessel of stainless steel or china, which rotates around its axis. On the ball mill case is a lockable opening, which serves to fill and empty the device. The internal volume of the mill is filled to a third with mill base and to another third with balls made of stainless steel, china or other hard materials. During the dispersion, the balls are seized by the rotation movement of the mill and are partly transported on high along the inside of the mill, until at a certain high they roll or glide back. Doing so they transfer impact and shear forces onto the pigment agglomerates, contained in the mill base. The advantages of the ball mill are the low emission (closed system) and the very good dispersing effect also to pigments, which are heavy to disperse. Due to the very long time of dispersion (as a rule between two days and one week) and its discontinuous production method, ball mills today are only rarely employed. Conversely, they still have a significant importance in the production of aluminium bronzes.

barium sulphate: Important filler (extender) with large field of application. On the market there are naturally extracted products (barytes) as well as synthetically produced ones (blanc fixe). In particular the blanc fixe types present an extensive particle range, showing ultra fine qualities, in addition to offering high lightness values. In combination with platelet shaped fillers like talc or china clay, the compact particle shape of barium sulphate makes it possible to achieve an optimal packing density in putties, fillers, dispersion paints and synthetic plasters. The fine types are also employed for price reduction in topcoats.

barrel coating: It is suitable to coat small mass production articles, if the demand referring to the uniformity of the coated surface is not too high. The small objects are put into a revolving barrel, similar to a basket, and poured with coating. After a short phase of dripping down, the surplus coating is mostly centrifuged. The air drying or stoving then takes place under the steady revolving of the barrel, in order to make sure that the objects don't stick to each other.

barrier effect: Effect, caused by platelet shaped fillers (extenders), which is above all positively noticeable in the corrosion protection. The platelets lie upon similar scales and in the film they form a mechanical barrier against the penetration of moisture, salt solutions and other corrosion agents.

baryte: → *barium sulphate*

base material: Term for the main component of two-component systems (two-pack systems). In addition to the main film former, the base material also contains all pigments and additives. The second system component is generally called hardener solution or cross-linker and it contains the cross-linking partner for the main film former (→ *two-component systems*).

bases: Chemical compounds, which in water solutions are capable of generating negatively charged OH-ions (→ *anion*). In their chemical behaviour they represent the counterpart of the acids and can be neutralised by the latter. Particularly important bases for the coating industry are the amines, as indispensable partners for the manufacturing of waterborne coatings (→ *amines*).

bead mills: It is a somewhat inaccurate, yet currently used general term for "attrition mills" (→ *mills*). That is, dispersing units which are designed according to the following principle: in a closed grinding compartment, filled with grinding media (beads), a rotor turns and puts into a rotating movement the mill base, which is pressed into the grinding vessel by means of a pump from the bottom or from behind the device in the case of horizontal systems. Thus, the pigment → *agglomerates* are seized by the also rotating beads and divided into primary particles and smaller aggregates by impactor- or shear-forces. When the mill base has reached the end of the grinding vessel, it leaves the closed grinding compartment through narrow separating slits, which hold back the grinding media. The grinding compartment is always equipped with a cooling jacket, which makes possible the circulation of cooling water for the purpose of temperature control. In the case of cylindrical mills, stirrers are mostly employed, which have several stirring disks at different heights. The conical nip-ball mills include rotors, which have as well a particular conical shape. As grinding media, beads of zirconium or aluminium oxide with diameters between 0,3 and 4 mm are mostly used, because of their hardness and high density. Meanwhile the closed bead mills are, beside the → *dissolvers*, the most commonly employed dispersing units in the coating technology. They can operate as single as well as cascade (series) of mills, which make possible several grinding passages of any amount of mill base at a very small solvent emission. By changing the dispersing → *parameters* it is possible to achieve different dispersion degrees. Due to their continuous mode of operation, low emission and high flexibility, bead mills have almost completely replaced in the course of time other dispersing units like → *triple roll mill* and → *ball mill.*

Bénard's cells: Characteristic honey-comb structure, which can occur on the surface of drying coatings due to floating problems (→ *floating*). Caused by the flow up and sink down of the streams in the liquid coating film, which promote the separation of the pigment particles, the edges of the cells are coloured differently, away from their centre. Thus, the colour of the coating surface does not appear uniform, which is naturally unacceptable.

bentonite: Phyllosilicate of natural origin, which is used as a raw material in the manufacturing of rheological additives (→ *thixotropie*). The mineral is composed of thin platelets, which are piled upon each other. Through a special after-treatment and an adequate delamination by means of suitable dispersion processes (→ *bentonite paste*), the platelets are enabled to generate a "house of cards" structure. Due to its bulkiness, this causes an increase of viscosity in the resting coating material.

bentonite paste: Semi-finished product of phyllosilicates (→ *bentonite*) and suitable solvents. During the manufacturing of bentonite paste the phyllosilicate is first delaminated by means of a non-polar solvent (mostly an aromate) and the employment of a dissolver, so that the piles of platelets are sheared apart. After that a polar solvent (generally an alcohol) is added. This activates the forming of hydrogen bridges and consequently of the thixotropy. The supple paste, which is generated in this manner, can be worked into different coating materials as a rheology additive, without specks, by means of a dissolver.

Berlin blue: Chemically a complex compound of iron, Berlin blue was the first synthetically produced, inorganic blue pigment. In use since the 18th century, it has been dealt under different names (Milori blue, Prussian blue, ironcyanogen blue). It is of major importance in the printing ink and artist's colour area industry. Instead, in the coating technology it is employed almost exclusively in very dark blue colours, because of its bad alkali stability and mediocre light fastness in strong blends with white pigments. In the blue colours referred to, it replaces phthalocyanine blue, which is inclined to → *bronzing*. For water-borne coatings Berlin blue is not suitable, because of the presence of alkaline amines.

binder requirement: Amount of binder, which is needed to produce a paste with defined flow properties, starting from a specified weight of pigment or filler (extender). The binder requirement can be numerically defined, according to the used test method. Most frequently it is determined as → *oil absorption value*, although the determining of the yield point or the Daniel titration with different binders are other methods used in practice. The binder requirement has major importance for mill-base formulations (→ *mill-base*).

binders: This term includes the total of the non-volatile components of a coating material (without pigments and fillers), which, during its processing, are responsible for the formation of the coating film. Their essential components are the film formers. As a rule, binders are synthetically produced, partly with natural raw materials modified polymers like alkyd resins, acrylates, isocyanates etc., which are employed in the form of solutions, dispersions or even of solid substances (→ *powder coatings*). Among all coating components, the binders influence most the range of properties of a coating material, because they are primarily responsible for the coating material's weather resistance, chemical stability as well as mechanical resistance.

biocides: Additives, which can prevent or reduce the microbial attack against coating materials and/or substrates. According to their effect, they are classified as preservatives against bacteria, fungi and algae. They can be used to protect the coating materials during their storage (in-can preservation) or for anti-microbial reinforcing of films (in-film preservation). A special role is played by wood preservatives, which prevent the biological degradation of wood due to soil contact. Unlike the past, when for such purposes harmful substances like high chlorinated compounds were also employed, today only physiologically inoffensive additives are used.

black pigments: → *iron oxide pigments, carbon black* and *spinel pigments*

blanc fixe: → *barium sulphate*

blazing of plastic surfaces: Pre-treatment method to improve the coating adhesion on plastic substrates. Plastics are normally too non-polar, in order to be optimally wetted by coatings. Through the treatment with an oxidising flame, that is with one which shows an excess of air, polar groups like ketones, OH-functions etc. are imposed on the plastic surface. Thus the polarity increases and the effect of the adhesion forces between the coating film and the substrate is improved (→ *adhesion*).

bleeding: Discolouration of a recently applied coating layer, caused by dissolution of a pigment, which is embedded in the underlayer. The reason for bleeding is the insufficient solvent resistance of some organic pigments. Particularly evident is this phenomenon in regard to simple azo-pigments. The term "bleeding" can be referred to red azo-pigments of the first generation, which, by recoating with a white coating material, generate in the white upper layer an increasingly reddish discoloration, as if the underlayer were "bleeding". Today this term is also used for other colour ranges.

blistering: Unwanted appearance in coating films, which may have different causes. Bubbles are often stirred in the coating material, during its production. However, they do not necessarily disturb the coating process, if they float spontaneously to the coating surface and burst there. This process can be made easier by application of special additives like degassing agents and defoamers. More problematic are the blisters, which arise during coating application and drying. These are mostly caused by formulation or process errors. In this regard, too fast an initial drying of the coating surface takes place, for example as a result of too high stoving temperature or of lack of high boiling solvents, so that moisture, inclusions of solvents or separated products, which are embedded in the film, cannot leave it in time. Thus they generate permanent blistering.

blocked, chemically: Expression used for hardeners, based upon isocyanates, amines or catalysts on the base of para-toluene sulfonic acid, which have lost their reactivity or accelerator effect at room temperature by a pre-reaction with special compounds. Therefore, the substances employed for such purposes are also called blocking agents

blocked isocyanates: Isocyanates are generally used as so called "hardeners" in → *two-pack systems.* If a one-pack material is needed which can be industrially processed, for example as electrical insulating material, packaging coating, powder coating etc., in this case the isocyanate must be "blocked" through a preliminary reaction with a suitable compound. As a consequence it is not able to react any more at room temperature. Only after its application, through the high stoving temperatures, a separation of the blocking agent takes place and the isocyanate gets back its original reactivity and can cross-link again the coating material.

blocking resistance: Printing resistance, above all of printing inks, packaging coatings, coil coatings, dispersion paints etc. Just a short time after their drying, these coatings have to be stackable without showing any print traces.

blown oil: Thermally refinished, oxidative drying oil with improved drying properties and an advantageous viscosity behaviour (→ *stand oil*). Unlike stand oil, in the case of the boiled oil the oxidative drying is partly anticipated by the addition (blowing) of air at 100 to 150 °C.

blue stain inhibition: Employment of special biocide preservatives in wood primers, above all for outdoor applications. These additives prevent or reduce the effect of wood decomposing fungi, which manifests itself as a greybluish discolouration, mostly on bright, soft woods like fir and spruce.

blue-yellow value: Index, used by the hue measuring of almost white or achromatic specimen, according to DIN 55980. Primarily it is taken into consideration in order to evaluate the whiteness of white pigments or fillers (extenders).

blushing: Typical incompatibility phenomenon of quickly drying coating systems, above all those based upon cellulose nitrate if, during their application, there are problematic temperature and moisture conditions (→ *incompatibility* and *dew point*). Caused by the evaporation cooling of the → *low boiling solvents*, moisture from the air can condense on the already wet film and form there a water layer, which generates a cloudiness called blushing at the interface between air and coating. This undesired phenomenon can normally be avoided by the addition of a solvent (→ *butanol*), which is capable of forming with the water a mixture which evaporates with constant composition (azeotropic) and in so doing it drags the water out of the system.

boiled oil: Term for binder solutions used in the printing inks sector. They contain mainly oxidative drying oils and high boiling mineral oils, but also other so called "sinking oils", able to penetrate into the absorbent substrate.

boiling: → *oil boiling*

boiling temperature: It is the temperature, at which the vapour pressure of a liquid reaches the value of the external pressure, which weighs upon the liquid surface. At this temperature the transition from the liquid to the gaseous state of aggregation does not take place only on the liquid surface (evaporation) but also within the whole liquid (boiling). Characteristic of the reaching of the boiling temperature is the formation of vapour bubbles in the deeper layers of the liquid. The boiling temperature is a constant value for each liquid substance, depending only on its surface pressure. If the pressure on the liquid surface increases, the boiling temperature increases too and conversely. Thus, in engineering it is possible to carry out different kinds of distillation, aiming at thermal liquid separation (vacuum and pressure rectification).

breathing protection: Catalogue of measures to prevent inhalation of dusts, solvent vapours or aerosols, during the coating production and processing. Among the measures are the extraction and ventilation equipment as well as the use of dust and gas masks with different filter inlays.

bronzing: Iridescent colour change, depending on the angle of vision. This phenomenon is based on the fact that the different crystalline nature of particles, belonging to the same pigment, can lead to a modified absorption behaviour along the different crystal faces. Bronzing is made very pronounced by many phthalocyanine blue pigments in full colour pigmentations, that is in very dark blue colours. In this case they phthalocyanine blue is often replaced by ironcyanogen blue (→ *Berlin blue*).

brush: It is the oldest coating application tool in human history. Its bristles can have different origin according to its mode of employment (horsehair, pig bristles, synthetic bristles). Also the shape can differ according to the application purpose (for example, round and flat brush, angular radiator brush). In spite of its long history, the brush is popular because it shows some outstanding advantages compared with other application tools: low price, ability to be used anywhere, easy transport and cleaning, absence of → *overspray*, possibility to coat also objects with complicated shapes. Naturally, the brush has also disadvantages: its non-uniform film-thickness, its rather insufficient levelling of the surface caused by the brush marks, and the manual labour it requires, above all in the case of large surfaces.

brushing: Manual coating technique, which utilises different variants of brushes (→ *brush*). A precondition for the applicability of coatings by brushing is a not too quick physical drying. Above all, decorative paints based upon long oil alkyds (→ *alkyd resins, long oil*), but also water borne acrylic glazes, show an ideal brushing behaviour.

bulking value: It is the volume of powdery substances, if they are only piled up without subsequent compression. In the coating technology, the bulking value plays a role in relation to the storage of pigments and fillers (extenders) as well as to the storage and processing of powder coatings.

butanol: Important solvent, belonging to the group of the → *alcohols*. Besides the polar properties, on which its partial solubility in water is based, butanol also disposes of a wide compatibility with non-polar compounds. Main solvent for amino and phenolic resins, butanol is used in various alkyd resin systems and water-borne coatings as well. A special property of butanol is its positive effect against haze formation, during the drying of nitrocellulose topcoats (→ *blushing*).

butyl acetate: It is one of the most important solvents for the coating technology. As an ester, it disposes of a wide dissolving power for the polar as well as for the non-polar range. Recently its importance has greatly increased, mainly because of the replacement of xylene (→ *aromatic-free*). Butyl acetate is the main solvent for isocyanate based two-pack systems.

butyl glycol: Water-soluble solvent with very positive properties as a film-forming aid for water-thinnable coatings and dispersion paints (→ *coalescing agents*), indispensable for all kinds of water-borne coatings.

C

cadmium-free: Largely fulfilled requirement, concerning colour formulations in the yellow and red range. The high danger potential to environment and health, caused by the heavy metal cadmium, was dealt with several years ago by the removal of cadmium pigments from the coating area. Therefore, their employment in coating materials is officially forbidden (→ *cadmium pigments*).

cadmium pigments: Inorganic pigments with particularly brilliant shades, good hiding power and outstanding heat resistance. Cadmium pigments are cadmium sulphide/selenide mixed compounds. Depending on their composition, their hue can vary from lemon yellow to bluish red. The main problem for this pigment group is the presence of the toxic heavy metal cadmium, in a form due to be easily dissolved in the gastric acid. Therefore, it can be assimilated by the human body as a result of inhalation of pigment dust, aerosols or sanding dust. According to an EU-directive the use of cadmium pigments in coating materials has been forbidden since 1996. Because of their high heat stability cadmium pigments have continued to play an important part in the plastic and ceramic industry.

calcium carbonate: It is the most commonly employed filler (extender) in the coating technology, above all due to its low cost. It is usually derived from natural mineral deposits but it can also be synthetically produced by precipitation. In nature, it appears amorphous as chalk as well as crystalline as calcite. Its preparation takes place by dry or wet grinding, according to the particle size range. As common properties, all calcium carbonates show a compact particle shape (advantageous for a high packing density), in addition to low hardness but also poor acid stability. However, relative to acid stability there are gradual differences. Crystalline calcite types are more acid resistant than the amorphous chalks. Even better, with regard to this property, are the mixed salts of calcium and magnesium carbonates, called dolomites, which are even harder. If particularly fine and bright calcium carbonate variants are needed, for example for partial replacement of the more expensive titanium dioxide, the precipitated types are the most appropriate. Calcium carbonates are main fillers (extenders) for dispersion paints. In this regard, for interior wall paints chalks are more often employed, while for exterior paints the calcite types or dolomites with higher acid resistance are more often used. Furthermore, calcium carbonates are used in all kinds of surfacers and primers, with the exception of application sectors where stability against acid rain plays a predominant role, like in the automotive industry and in high performance coil coatings for building protection.

camouflage paints: Paints, which are employed to mask military installations as well as battle armament of soldiers and other war devices. Since in the modern warfare the unveiling doesn't take place by means of the naked eye but of infrared (IR) or ultraviolet (UV) visual display units (the latter by snow camouflage), camouflage paints have to be pigmented in such a way that they can reproduce the reflection curve of the chlorophyll in the IR-range and that of the snow in the UV-range. For the IR-camouflage, cobalt pigments are most often employed (→ *spinel pigments*). The reflection curve of snow in UV-camouflage can only be reproduced satisfactorily by using white lead, which has to be labelled, because of its toxicity.

can coating lacquers: Coatings, fulfilling the purpose to protect permanently the can metal against the influences of aggressive components of its food contents. At the same time, they have to prevent metal traces or corrosion products from migrating into the food. Also, in very thin layers can coatings lacquers have to be non-porous, chemically resistant and suitable to sterilisation. Besides, they must not cause change of taste and aroma in the food contents.

carbon black: It is by far the most important black pigment. Actually it consists of pure micro crystalline carbon, which is produced through combustion of hydrocarbons by deficiency of oxygen and subsequent immediate cooling. Because of its low price, its outstanding weather and chemical resistance and the high tinting strength, it is commonly used in all fields of the coating technology. The only exception are coating materials, whose dispersion take place by means of dissolvers like dispersion paints, floor varnishes, primers and surfacers. Due to its production process, carbon black can be originated with different primary particle sizes (→ *primary particle*). The "lamp black process" produces the coarsest types (particle sizes between 200 and 50µm). Through the "furnace black process", pigments with particle sizes between 70 and 15µm are originated. The finest pigments, and therefore also the most difficult to disperse, are obtained by the "channel black process" (particle sizes between 30 and 10µm). The channel black pigments are subjected to an oxidation treatment in order to get a better wetting of the pigment surface and, as a consequence, facilitate considerably the dispersing process. According to their particle sizes, the carbon black pigments differ from each other in their appearance. A lamp black is in its full colour, because of its scattering power, much more greyish than a channel black. In combination with scattering or reflecting pigments, like titanium dioxide or aluminium bronze, a lamp black produces more bluish grey hues than a channel black. Also important is the electrical conductivity of carbon black pigments, which is used for manufacturing so called conductive coatings, to improve the electrostatic application on plastic substrates.

carboxyl group: Functional group of the carboxylic acids (-COOH). It has a decisive importance concerning the synthesis of many film formers, because it is able to bring about an esterification reaction with the hydroxyl groups of the alcohols (→ *esterification*). Thus the carboxyl groups are involved in the formation of all types of polyesters (→ *alkyd resins, polyester resins and polyester resins, unsaturated*) as well as of the → *acrylic resins*. In the coating analysis and in the binder synthesis the → *acid value* is used to determine the presence of free carboxyl groups.

castor oil: Non-drying vegetable oil, whose main fatty acid component is the → *ricinolic acid*. Castor oil is principally used to manufacture short oil alkyd resins for stoving systems and nitrocellulose combination lacquers (→ *alkyd resins, short oil*). Because of the presence of an OH-group in the molecule of the ricinolic acid, castor oil shows a particularly good compatibility with nitrocellulose and amino resins. Besides, it is employed as raw material for the production of dehydrated castor oil (→ *castor oil, dehydrated*).

castor oil, dehydrated: It is produced by thermocatalytic water separation from the natural castor oil. Through the water loss, the non-drying → *ricinolic acid* is turned into a semi-drying fatty acid with two conjugated double bonds. This variant of → *linoleic acid* is contained for almost 30 % in the technical dehydrated castor oil. The dehydrated castor oil is employed for the production of oxidative drying as well as for stoving alkyd resins with subsequent oxidative cross-linking (→ *alkyd resins, medium oil*)

catalyst primer: Also called "reaction primer". Priming coat, containing peroxide, which is used in a rather antiquated process for wood coating with unsaturated polyesters. In this, also known as "sandwich process" technology, first the catalyst primer and then the unsaturated polyester are applied onto the wood by means of a curtain coater. At the contact surface between the primer and the polyester layers, the peroxide starts the radical polymerisation, which causes the curing of the unsaturated polyester resin.

catalysts: Substances, which are able to accelerate a chemical reaction, without altering themselves. In the coating area, there are a lot of important examples of catalysts. In the first place, the metal soaps should be mentioned, which are used to accelerate the oxidative drying of alkyd resins (→ *driers*) but also the polymerisation of unsaturated polyesters (→ *polyester resins, unsaturated*). Worth mentioning are also acids for the so called "acid curing coating", tin compounds for aliphatic isocyanates and amines for different coating systems. Not to be confused with the catalysts are the "initiators", for example organic peroxides, which are able to start a radical polymerisation in the form of so called "hardeners" of polyester putties, by decomposing themselves into radicals.

cathode: Negatively charged electrode, to which the positively charged cations in a electrolysis bath migrate to discharge, by application of a direct voltage. This principle is the basis for the coating deposition the cathodic dipping process.

cathodic electrodeposition coating: → *electrodeposition coating, cathodic*

cation: Particle, which is electrically positively charged. It consists of either a single atom or a group of different atoms. In the coating area, this term is narrowly associated with the cathodic electrodeposition coating (→ *electrodeposition coating, cathodic*).

cellulose acetobutyrate (CAB): Mixed ester of cellulose and organic acids, with outstanding light fastness and heat resistance. It is mainly employed as a physically drying binder, although sometimes it can also be chemically cured by means of its OH-group content. Its by far most important field of employment are solvent-borne metallic base coats for automotive two-layer coatings. There, as a result of its quick drying, CAB fixes the aluminium flakes and simultaneously encourages their parallel alignment because of its volume shrinking during the drying. Thus the metallic effect may be achieved in an optimal way. The good light fastness in this case is absolutely necessary because of the particularly high ultra violet light dose, caused by the mirror effect of the aluminium platelets.

cellulose nitrate: Ester of cellulose and nitric acid, also known respectively as "nitrocellulose" and "collodion cotton". Cellulose nitrate is the oldest among all cellulose derivatives, which are used as film formers. The nitration of cellulose is necessary to obtain sufficient solubility in organic solvents. In this regard the types of cellulose with a low esterification degree - expressed as the nitrogen content - are principally soluble in alcohols (A grades) while those with a high esterification degree are soluble in esters (E grades). The solutions are characterised by a very quick physical drying. Even though the importance of nitrocellulose as a film former has been decreasing markedly in the last decades, primarily because of the bad light fastness and chemical resistance, nevertheless considerable amounts of collodion cotton are still employed, often in combination with non-drying alkyd resins (→ *nitrocellullose combination lacquers*). The main fields of application are quick-drying coatings for wood and furniture, industrial coatings, fillers and putties.

cement resistance: Important requirement for all paints and their components, which are used to coat concrete, brickwork and similar building materials. Relating to this, the resistance against alkaline solutions (→ *pH-value*), which are generated by the impact of moisture on such substrates, plays a main role. Thus, saponifiable film formers and additives must not be employed in such systems (→ *saponifiability*). Likewise the use of pigments, which are labile against alkali exposition, for example Berlin blue, is not advisable.

chalk: → *calcium carbonate*

chalking: → *photochemical degradation*

channel carbon black: With a particle size range between 10 and 30nm, the channel carbon black has the finest particles of all carbon black variants but also the most disadvantageous dispersing properties. In its production process, highly boiling oils rich in aromates are evaporated, and in a mixture with a combustible carrier gas they are led to burners which develop many little flames. The combustion gases are chilled by a cooling cylinder and so the carbon black particles are formed. To improve the dispersing properties, the channel carbon black is subsequently oxidised. In blends with white pigments or aluminium bronzes it generates significantly more yellowish grey colours than the → *furnace carbon black*. In its full-tone (hiding), the channel carbon black is deep black.

chemical resistance: Resistance of coating materials against acids and lyes, as well as against other aggressive substances like oxidation and reduction agents, food components, preservatives, detergents, grease etc.

china clay: → *kaolin*

chlorinated hydrocarbons: Group of solvents with outstanding dissolving power, which nevertheless are not used any longer in coating formulations, because of their high risk potential for health (liver toxicity) and environment (ozone hole). Only methylene chloride represents an exception because of its employment in paint remover pastes, which however can be considered in decline.

chlorinated paraffins: Chlorinated, chain-like hydrocarbons with a chlorine content of 50 to 70 %. This makes them incombustible. Chlorinated paraffins are mostly used as chemically resistant plasticisers, above all in chlorinated rubber and polymers with high requirements relative to the fire protection properties of the coating material.

chlorinated rubber: Originally a derivative of natural rubber, chlorinated rubber is now mainly produced by chlorination of synthetic polyisopren. Chlorinated rubber is an exclusively physically drying binder with only mediocre heat stability (not more than 80 °C continuous heat exposition). On the other hand, it shows an outstanding water and chemical resistance and it is hardly combustible. Its main fields of employment are anti-corrosive primers and chemically resistant coatings, above all for underwater construction, ships, manufacturing plants etc.

chromate-free: In chromate pigments the metal chromium is present in its valence VI, which is, as it has been proved, carcinogenic (unlike these, compounds of the trivalent chromium are physiologically harmless → *chromium oxide green*). Therefore, many efforts have been made for several years to eliminate chromate from coating formulations. Already in the eighties pigments like zinc and strontium chromate, which are relatively easily soluble in the gastric acid and therefore particularly dangerous, were preponderantly replaced by harmless phosphates. Today, the replacement of the low soluble chromium yellows and molybdate reds is often required, due also to their lead problems (→ *lead chromate* and *lead molybdate*).

chromaticity diagram: Due to its form also called "colour triangle" or "colour pyramid". It is described in the DIN 6164, part 1 and represents the basis onto which the DIN colour system is constructed and from which other colour systems (for example CIE L*a*b*) derive as a result of mathematical transformation. The chromaticity diagram is represented in a two coordinates system, where the colours are disposed in accord with their standard colour coordinates x and y. The boundary corresponds to the spectral colour line, that is the sequence of the hues with maximum saturation (highest intensity, which can be reached only with spectral lights). In the middle of the chromaticity diagram, there is the achromatic point for the standard illuminant D 65 (→ *illuminant*). Some times, the chromaticity diagram is additionally illustrated by a brightness axis, which cuts through perpendicularly the achromatic point. A line, connecting a point of the boundary with the achromatic point, contains only colours with the same hue (T) but different saturation (S). The third factor, which completes the colour characterisation, is the darkness degree (D).

chromating: Pre-treatment method, especially for substrates of aluminium and light metal alloys. By means of chromate containing solutions, so called "conversion coatings" are formed on the metal surface. These can be green or yellow coloured, depending on the composition of the chromating agent. Such conversion coatings generate an ideal adhesion between the metal substrate and the applied coating layers.

chromium oxide green: Oxide of the trivalent chromium and therefore physiologically harmless (→ *chromate-free*). This inorganic green pigment is characterised by high hardness, outstanding chemical and heat resistance and a low price. However, its shade is rather dirty and so it cannot be used to formulate brilliant green hues. Its main application fields are industrial coatings and floor paints with high abrasion resistance, chemically resistant coatings and dispersion paints.

chromium titanium yellow: Inorganic yellow pigment, which is increasingly used to replace lead chromate in lead-free pigmentations (→ *lead-free*). Its relatively dull, reddish yellow shade does not make possible any direct substitution of lead chromate. However, in comparison with iron oxide yellow, its hue can be considered rather pure and therefore more suitable to match brilliant, reddish yellow shades in combinations with organic pigments. Like → *nickel titanium yellow*, chromium titanium yellow is based upon a rutile host lattice with incorporated external ions, in this case chromium ions, with colouring effects. Also its heat stability and therefore the suitability to employment in coil coating systems is comparable to that of nickel titanium yellow. However the hiding power of chromium titanium yellow is considerably higher.

CIE L*a*b*-system: It is the most widely used colour system in the coating technology. The → *tristimulus values* of the colour test panel, which are determined by means of a spectrophotometer, are inserted using a mathematically converted form, in a three coordinates system, which describes the CIE L*a*b* uniform colour space. There, the axis a (for red/green) and the axis b (for blue/yellow) intersect each other in a plane. The brightness axis L passes perpendicular to the intersection point of a and b (achromatic point). The values of the three coordinates characterise a definite point in this colour space. By measuring a coloured sample against a reference (standard), the distance between two points in the colour space can be mathematically determined as colour difference (→ *colour measurement*).

circulating air oven: → *convection drier*

citric acid esters: Plasticisers for coating systems, which are used in the food and cosmetics sector and hence must show physiological harmlessness.

clear coats: Unpigmented and therefore transparent finishing enamels, which have a protective function for the layers below but without changing or hiding their appearance. Clear coats can be one-layer systems, which are applied directly on a substrate, whose optical characteristics have to be conserved (for example clear coats for wood, paper or building materials); they can also represent the upper film in a multi-layer system.

closed-cell: Current term to specify wood paints (→ *open-cell*). Wood coatings are called closed-cell, if they form an impermeable and therefore "nonbreathing" film. Since in wood parts coated in this manner moisture interchange is not possible any longer, the wood has to be stored and dried sufficiently before the coating process. Closed-cell coatings are preferably used for objects that must not change their form or appearance even after a long exposition to water, chemical and detergent agents or that have to show high mechanical resistance like, for example, parquet flooring, wood stairs, furniture, toys etc.

clouding: Defective appearance, concerning the application of metallics, which depends on local differences created by the orientation of the aluminium platelets within the coated area (→ *two-coat metallics*). It may be the result of an incorrect application, if for example the metallic coating is sprayed too wet so that the fixing of the aluminium platelets and the shrinking of the film, which are both indispensable for their horizontal orientation, do not take place to a sufficient degree. It can also be caused by formulation mistakes, like a non-balanced solvent composition or the absence of special additives like polymer waxes.

coagulation (of an emulsion): Also called "breaking of the emulsion" (→ *emulsion*). In this connection, the in water dispersed liquid polymer particles flow together and form coarser particles, which are not able to float any more. Depending on their density, the particles are inclined to settling or creaming and form with the water a two-phase system. The causes of coagulation are multiple and depend, among other things, on the chemical structure of the emulsion. The most important reasons of coagulation are: too high temperature, the presence of external ions and an incorrect pH-value. However, in the → *electrodeposition coating*, the film-forming on the work pieces is provided by a forced coagulation of the coating material, by means of an electrolysis.

coalescence agents: Also called "film-forming additives". They are organic solvents, such as butyl glycol, which show a good water miscibility and are added to aqueous film former dispersions in little amounts, in order to facilitate their film-forming. In the last stage of the film-forming process of such dispersions, when the water by evaporation has almost completely left the coating material, the dispersion particles have to flow into each other or to stick together, in order to form a closed film. This process is highly dependent on the viscosity of the particles and therefore on the application temperature. If the application temperature falls below the so called "MFT" (→ *minimum film-forming temperature*), the flow properties of the particles are reduced in such a way that a successful film-forming is not possible any longer. By employment of coalescence agents, a decreasing of the MFT takes place, so that also by temperatures up to 5 °C an optimal film-forming is guaranteed.

coating formulation: → *formulation*

coating materials: Common term for all products of the coating and paint industry. Beside the finishing coatings and paints it refers also to other materials of the coating technology like putties, fillers, primers etc.

coating of plastics: The coating application on plastic substrates takes place, as a rule, by means of pneumatic atomisation (perhaps with electrostatic assistance) or of → *rapid-rotation atomisers*. By the electrostatic method, a sufficient conductivity of the plastic parts has to be achieved. Since the plastic materials are rather bad conductors, they are either equipped with conductive pigments or coated with conductive primers. Before application, a careful pre-treatment of the substrate is necessary. All remains of release agents, as well as grease or sweat, must be completely eliminated by "suitable" solvents.

coating production: The process chain of coating manufacturing comprehends several steps, which can be mostly identified with the name of typical stations in the coating plants. Following, the purposes of these stages are briefly described. "Row material warehouse": storage of the row materials and semi-finished products and their supply to production – "Mill-base production": preparation of master batches (pigment concentrates) and their homogenisation and pre-dispersion by means of dissolvers – "Dispersion": main dispersion process of the mill-base, as a rule by means of attrition mills – "Let-down": completion of the mill-base by addition of the remaining formulation components, also called "cutting back" or "making up" – "Tinting": matching of the required hue – "Packaging": purification of the coating material by screening/filtration and filling in the designated containers.

coatings for plastics: Coating materials for objects of plastic. The coating of plastic workpieces not only takes place for aesthetic reasons, but also in order to protect the technological properties of the plastic materials against damaging influences (for example, UV-radiation, fuel, and dirt.). Besides, structured and so called "soft-feeling coatings", able to confer plastics a somewhat more comfortable feel of soft surface, for example for the interior equipment of cars or for the furniture sector, become more and more popular. To fulfil these functions, coatings for plastics have first of all to show good and stable adhesion on the plastic substrate. In the coating formulations this property plays an outstanding role, because most plastics are rather problematic substrates in terms of adhesion, due to their low polarity. Besides, they are more or less susceptible to solvents. In the coating formulations, additives and solvents have to be chosen very carefully, according to the specific properties of the plastic substrate, in order to provide a polarity balance without showing a too aggressive dissolving power to the substrate. In the latter case, there may be the risk of formation of tension cracks (particularly for polycarbonate). Often sufficient adhesion can only be achieved by means of a special substrate pre-treatment, aiming to increase the plastics polarity.

coating systems: Term, which describes the combinations of different coating materials. Together they form multi-layer coats, according to the recommendations of the coating manufacturer. In the multi-layer coat, the single components of the system fulfil specific tasks, which are carefully coordinated one with the other.

cobalt blue: Mixed-phase oxide from bivalent cobalt and trivalent aluminium (cobalt aluminium spinel), with incorporated chromium ions. By increasing the chromium amount, the pigment colour becomes more and more greenish. Cobalt blue pigments have an outstanding chemical resistance and can abide temperatures of up to 1000 °C. Since they are relatively expensive and also show a rather low tinting strength, in the coating technology cobalt blue pigments are only used in special application fields (for example camouflage paints, heat resisting lacquers etc.).

cobalt drier: The most important additive to promote the surface drying of oxidative drying oils or alkyd resins (→ *driers*). Due to the ability of cobalt to change its valence (between two and three), this drier ranks among the "primary driers", these are able to accelerate the radical forming and therefore the incorporation of atmospheric oxygen, during the drying process.

coconut oil: Non-drying oil of natural origin. Due to its low yellowing tendency and its cheapness, it is often used to produce non-drying alkyd resins for stoving systems and as plasticisers for physically drying film formers like cellulose nitrate and chlorinated rubber.

cohesion: Interaction of physical and chemical forces of attraction, which are responsible for the strength of the internal coherence of a substance, for example within a coating layer.

coil coating: Method for the continuous coating of metal strips. At the beginning, the strip is available as a coil, which is uninterrupted decoiled, during the whole process. In a continuous coating line, the strip passes at high velocity the stages of cleaning, phosphating, priming, coating, stoving, and if necessary laminating. At the end of the process, the coated strip is recoiled. The coating application is carried out by means of rollers and the stoving process takes place in a closed tunnel oven with thermal after burning for the solvent (non-polluting). The coil-coated metal sheets are supplied to the industry, where they are processed to parts for refrigerators, washing machines, commercial vehicles etc., by means of blanking, bending and drawing. The coating layer has to endure these deformations without damages.

cold check test: Wood is a "warping material", that means, it changes permanently its form by swelling and contracting, due to moisture and temperature effects. This form change depends strongly on the type of wood, but also on the kind of fibres (heartwood or sapwood). Additionally, it depends on the fibre direction (anisotropy of wood). The elasticity of wood paints for outdoor application must be coordinated with the volume variations of the wood substrates, in consideration of the loss of elasticity of the employed binder due to decreases in temperature. In this regard, a widespread test method in the wood coating industry is the so called "cold check test", where coated wood specimens are subjected to temperature shocks from below -20 °C up to over 60 °C.

cold resistance: Technological property, of coating materials, which first of all depends on the elasticity behaviour of the binder at low temperatures. With increasing cold, the elasticity of polymers decreases to a high degree. This must be taken into consideration by coating application, primarily in connection with "warping substrates" (→ *cold check test*), so that this loss of elasticity does not provoke coating damages like cracking or flaking.

collodion cotton: → *cellulose nitrate*

colloidal solution: Dispersed system of polymer particles in a solvent, where the diameters of the dissolved particles average out between 0,1 and 0,01µm. The colloidal solutions rank between the "true solutions" (particle size below 0,01µm) and the emulsions or suspensions (particle size above 0,1 µm). While true solutions are completely transparent and emulsions and dispersions milky, colloidal solutions show only a light opalescence and are almost clear.

colophony: The only natural resin, which plays a noteworthy role in the paint technology. It is extracted from pine trees, together with → *turpentine oil.* Depending on the method of extraction, two different resin types can be differentiated: the gum resin, obtained by slitting the trunk, and the wood resin, extracted from the rootstock. Colophony is also obtained by the extraction of the so called → *tall oil.* Colophony consists of different isomers, that means structural versions of the abietic acid. The compact structure explains the solid state of aggregation at room temperature and consequently its resin character. Before employment, colophony has to be "refined" by neutralisation of its acid groups. Colophony is used as brittle resin in physically as well as in oxidative drying paints.

colourant: Generic term for all colouring matters. Their classification can take place according to, respectively, technological properties (DIN 55949) or colour shade and chemical properties (DIN 55944). Dyestuffs and pigments represent the main colorant groups.

colour: Sensory perception, which arises by the interaction of electromagnetic radiation with wave-lengths between 380 and 750nm in the so called visible spectrum, and the biological sensors in the human eye. When visible light enters the eye, two different kinds of biological detectors are activated: the retinal taps and cones. While the retinal taps react upon the whole light intensity in a non-selective way, therefore they can only distinguish between dark and bright, the different types of retinal cones are selectively activated by red, blue and green light. According to the stimulation degree, electrical signals are formed and forwarded to the brain trough the optic nerve. By means of these signals, the brain "assembles" the colour.

colour, achromatic: Colour, which is exclusively characterised by its lightness value. Achromatic colours are White, Black and the whole Grey scale.

colour, chromatic: Colour, which is not only defined by its lightness value, but which additionally shows a characteristic hue like Red, Yellow, Blue etc.

colour circle: Circular system, suitable to order the colours according to their sensorial distance from the so called elementary colours Yellow, Blue, Red and Green, which in the colour circle stand opposite to each other and form the colour axes of the circle. The colour shades between them form the transition from an elementary colour to the next one. Colours, which in the colour circle are opposite to each other are → *complementary colours*. By matching colours, the colour circle makes easier to choose suitable tinting components. To successfully match brilliant colour shades for example, it is necessary to use only coatings or pastes, which in the colour circle are near each other.

colour contrast: Influence on the colour evaluation, caused by a chromatic outer field. A hue appears more saturated the more the colour of its environment approaches that of its own complementary colour. A red colour, for example, appears as a more intensive red in a green outer field than in a red one.

colour difference: Geometrical distance between two colour samples, as a rule a standard and a colour matching panel, in the uniform colour space (→ *CIE L*a*b* system*). The total colour difference is specified as ΔE-value.

coloured pigments: Colorants, which are not only defined by their lightness value, like white and black colours, but also by a characteristic shade (yellow, red, green etc.).

colour fastness: Very important property for all kinds of outdoor paints, particularly for aviation coatings. These must not change their colour in a significant degree even after a long period of weathering, otherwise in case of refinishing colour differences would become visible. The colour fastness depends basically on the light fastness of the employed pigments. However, yellowing and chalking of the binder may also have a lasting effect on the colour.

colour formulation calculation: Creation of colour formulations by means of colorimetry (→ *colour measurement*). In this connection, a computer program falls back on the stored colorimetric data of numerous pigments, respectively tinting bases or mixing coatings, and tries to mix them by means of calculations to form a reflection curve, which as much as possible coincides with the stored reference curve.

colour index (C.I.): Numerical system for the classification of colouring materials, according to their chemical composition. In the C.I., first the hue group of the pigment is indicated by two capital letters, which derive from the corresponding English terms (for example, P.Y. means Pigment Yellow); then a number follows, which identifies the chemical compound class. Products from different manufacturers with the same number are thus chemically identical. However, relative to the shades, they can show differences because beside the chemical composition the physical particle properties (particle size and shape) may play a significant role.

colour location: Position of a colour in a colorimetric uniform colour space (as a rule in the CIE L*a*b* coordinate system).

colour measurement: Method for the measurement and illustration of colours, by means of numbers, corresponding to the visual perception of the eye of a statistically so called "normal observer", which represents the average colour sensitivity of the human eye. A particularly important duty of colour measurement is the numerical description of colour differences. This represents the foundation of modern colour matching and colour formulation calculation (→ *CIE L*a*b* system*).

colour measurement instrument: Device for measuring the tristimulus values, that means the content of Red, Blue and Green in a colour. Using these values, the position of a colour in the uniform colour space can be determined and the colour difference between two different specimens can be calculated. Two principal measuring techniques are used. The three-colour process employs three sensors with filters, which correspond to the eye sensitivity and deliver directly the tristimulus values. The more complex, but also more exact, spectrometric method decomposes a standardised white light in single wavelength bands, which are sent one after the other to the specimen. A photo cell measures the amount of the corresponding wavelength, which is reflected by the specimen. From the reflection values of the single wavelengths, a reflection curve is formed, which reproduces the light absorption behaviour of the coloured specimen across the whole visible spectrum. From this curve, the computer calculates the tristimulus values with the aid of the sensitivity curves for the eye and the spectral composition of the employed white light. The modern spectrometers dispose of an extensive software, able to calculate, starting from the tristimulus values, the standard colour coordinates, colour differences between two specimen and colour formulations.

colour standard: Often also called reference chart or panel. It is a specimen of the shade, which the tinter has to obtain by matching a colour. The colour standard can be required by a customer, it can also be an internal reference specimen of the manufacturer or a RAL colour chart (→ *RAL register*).

colour value: Method to classify the colour of clear liquids. In the coating technology, it is mainly employed to evaluate the colours of binder solutions. The following two values are mostly used: the Gardner colour value (ISO 4630) and the Haze colour value (ISO 6271). According to the Gardner method, a detector tube with a specimen of the liquid, whose colour value has to be determined, is inserted in the so called Helligecomparator and compared with coloured glass standards. The Hazen method is based upon the comparison of a detector tube, containing the liquid specimen, with a series of detector tubes, which are filled with chloride/hexachloroplatinate solutions in different concentrations.

combustibility: Capability of a substance to react chemically with an oxidising agent, normally oxygen in the air, causing the formation of heat and light. The inflammability of numerous coating raw materials, above all solvents, is a high threat to the coating industry. The classification of the dangerousness of inflammable substances, according to different technical regulations, takes place in relation to the →*flash point* or to the →*ignition temperature.*

combustion: Chemical reaction between a combustible substance and an oxidising agent, mostly air oxygen, which takes place and develops heat and light. To start the combustion process, an ignition source (for example open flames or sparks, provoked by electrostatic charge) or the achievement of the →*ignition temperature* of the combustible material is necessary.

compatibility: Condition of mutual acceptance of raw materials in a mixture (→*incompatibility*). Compatibility is produced by a balanced coordination of the →*polarity* of the single mixture components and is a basic precondition for the stability of coatings. As it is known, the "mixture" named coating contains the most different raw materials, which by nature do not always offer the optimal conditions to achieve lasting stability. Their compatibility has to be created in part by means of surfactants (→*wetting*) and/or by employment of suitable solvent blends. Compatibility tests (storage tests) belong to the most important points on the checklist of the paint technologists.

complementary colours: Colours, which when are mixed together produce achromatic hues. The result of mixing complementary coloured lights (→*additive colour mixture*) is a white light. Conversely, in the case of pigments, complementary colours produce black or grey hues (→*subtractive colour mixture*). The latter result is very important for tinting processes in the painting technology. In the →*colour circle,* complementary colours are situated opposite each other. Tinters adhere to the rule that the greater the distance between the hues of two mixing colours in the colour circle, for example in the form of tinting pastes, the "dirtier" the hues of the mixture becomes, because of the increasingly "black content".

compound (chemical): Substance, which is formed through the combination of atoms of different chemical elements. The cohesion of these atoms is provided by chemical bonds, which, according to the respective atom combination, may be differently natured and, among other things, may influence the →*polarity* of the compound. The compounds are characterised by physical and chemical properties, which are different than those of the single elements. Above all, in organic chemistry it is possible to obtain extremely versatile properties by the variation of the atom combination.

compressed-air spraying: →*High* and *low pressure spraying.*

condensation resins: Film formers, which are synthetically produced and, if necessary, cross-linked by a polycondensation mechanism. The polycondensation presupposes that several low molecular components with reactive groups (functional groups) react with each other to form coarser molecules. At the same time small molecules of reaction products, like water, alcohols, formaldehyde etc., are separated. Typical polycondensation resins are saturated polyesters, alkyd resins, amino resins etc. Important components of condensation resins to be mentioned are: polyols (functional group = -OH/→ *hydroxyl group*) and polycarbon acids (functional group = -COOH/→ *carboxyl group*).

conditioned room: Room, where temperature and air moisture are permitted to vary only between determined values. Conditioned rooms in the coating technology are important primarily for testing and application. The reproducibility of many physical and technological coating tests is only guaranteed after a corresponding storage of the specimens in a conditioned room, because temperature and moisture can influence film properties like elasticity, hardness, conductivity etc.

contact angle measuring: Method to determine the interfacial tension between a solid surface and a liquid. In the coating technology this evaluation gives information about the actual wetting conditions (→ *wetting*) between film former solutions and pigments or between coatings and substrates to be wetted. According to this method, the angle is measured which is formed at the edge of a droplet, when the liquid is dropped onto the solid surface, between the tangent of the droplet curvature and the surface. The larger the angle, the bigger the tendency of the liquid to contract and to form droplets; the smaller the angle, the bigger the propensity of the liquid to spread on the surface and to wet it. Spontaneous wetting can only be expected if the contact angle is smaller than 90°. If it is larger, it is necessary to diminish the interfacial tension by the addition of wetting agents.

convection drier: Drying device, which uses heated air to transfer heat to the coated work piece. The air is heated by means of a heat exchanger and continuously carried to the object. In between, through solvent vapours enriched air is partly replaced with fresh air to avoid the forming of an explosive atmosphere. There are different types of convection driers like circulating air ovens or tunnel ovens, often combined with → *infrared drying.*

cooling by evaporation: During the evaporation process, the solvent molecules absorb energy, because in the gaseous state of aggregation they have a higher mobility than in the liquid one. They take this energy from their environment, which as a consequence is cooled down. This cooling is responsible for the forming of whirl-shaped convection streams in the drying coating film, because the upper liquid layers become heavier and sink down. This fact promotes the phenomenon of →*floating.* A further appearance, which is caused by the cooling through solvent evaporation, is the → *blushing* of fast drying coating films (→ *dew point*).

copolymerisation: Production of polymers by employment of different monomers (→ *polymerisation*). The producers of raw materials for coatings use this method to optimise the properties of film formers. Type and amount of the involved monomers make possible to design polymers with "tailor-made" property profiles.

copper phthalocyanines: Group of organic pigments, which cover the range from a yellow-tinted Green to a red-tinted Blue. They are characterised by outstanding purity and tinting strength and they also rank among the most light-, weather- and solvent-resistant organic pigments. These properties, in addition to rather moderate prices, make them all-purpose pigments for the whole coating technology, which can be used for plain colours as well as for metallic colours. In the blue range, the different shades are produced by different crystal structures. In the green range, the presence of chlorine and bromine atoms or the absence of copper are responsible for the colour location.

corona charge: Charging process of powder particles, during their electrostatic application (→ *electrostatic powder coating*). In this connection, air components are ionised in a high voltage field of 80 to 180KV by means of a negatively charged electrode, which sprays electrons. When leaving the spray gun, the powder particles can be electrically negatively charged either by direct attachment of free electrons or (predominantly) by electron exchange with the ionised air.

corona discharge: Pre-treatment method for plastic surfaces, aiming to increase their polarity and thus improve the coating adhesion. By means of an A. C. voltage electrode, a so called luminous discharge is generated, which forms high-energy oxygen ions. On their part, these ions oxidise the plastic surface and make it more polar.

corrosion: Chemical process, which leads gradually to material decomposition. For technological applications, the corrosion of metals like iron, zinc and aluminium, which represent very important alloy components, plays a main role. It is possible to distinguish between acid corrosion, → *moisture-induced corrosion* and corrosion through formation of so called "local cells", caused by the contact of two metals with different potentials. Corrosion protection represents one of the most important economic tasks of the coating technology.

corrosion protection, heavy duty: It refers to the protection of durable goods like all types of buildings, industrial installations, bridges, harbour facilities, ships etc. It is mostly a matter of thick coat systems, among other things containing coarse platelet shaped fillers (extenders) like micaceous iron oxide or mica, to obtain an optimal → *barrier effect*. As a anticorrosive pigment → *zinc dust* is generally employed.

corrosion protection, OEM, refinish, industrial etc.: It concerns relatively short-lived objects of utility like cars, trucks, architectural parts, gardening tools etc., which are normally used outdoor. In this context, only coating systems with rather low film thickness, like wash primers, → *corrosion protection primers*, and fill primers, are employed. Normally these contain fillers (extenders) with low particle size and zinc phosphate pigments against corrosion.

corrosion protection primers: → *primers*.

corrosion tests: The only absolutely reliable corrosion test is outdoor weathering under the climate conditions, which are relevant to the practical application of the corrosion protection systems (→ *outdoor weathering*). Actually, such long term tests are carried out for all corrosion protection systems, which play a significant role in the market, if a conclusive evaluation of their quality is required. However, during the development of such systems, the technicians depend at first on accelerated tests, able to give quality information in comparison with competitive products or standards. Thus, different corrosion tests with specific focal points have been developed. In the salt spray test to DIN 50021 or ISO 7253, the test panels are hung up in a closed chamber and, at elevated temperatures, continuously sprayed with a salt solution. Often the test panels are scratched before testing to evaluate the rust formation (sub-surface migration) on unprotected spots. When only the effects of temperature and moisture are required, as a rule either the saturated atmosphere test, which exists in different variations, or the water immersion test is chosen. The alternating atmosphere test to VDA 621-415, which is a combination of the salt spray and saturated atmosphere tests, is considered a particularly suitable corrosion test for automotive coatings and generally for zinc or galvanised steel.

co-solvents: Denomination for organic solvents, which are employed in water-thinnable coating systems. Their name makes clear that they fulfil the task of encouraging solubility. This may be detected as a reduction or elimination of the so called→ *water mountain*, which is a precondition for the processing of such systems. The co-solvents also play a positive role relating to the film forming phase, because they keep the film "open", during water evaporation, and hence they prevent the formation of blistering and pin holes. As "temporary plasticisers", they facilitate the film forming by promoting the smelting of the dispersion/ emulsion particles. Through the use of co-solvents which are able to form azeotropes with the water, water residues can also be entirely entrained out of the deeper layers of the film.

cracking: The causes of cracking in topcoat layers can be numerous. Cracking may derive from a wrong choice of material, for example if a high filled coating is applied onto a substrate which contains much film former (→ *crackling finish*) or if a coating material, with low elasticity, is used to coat a substrate, like wood, which "warps". Cracking may also be based on application defects, for example if a dispersion paint is processed under its MFT (→ *minimum film-forming temperature*) or if a putty is too thickly applied. However, the most frequent cause of cracking can be found in coating material degradation due to long weathering exposition. Above all, the effect of ultra violet radiation often provokes medium-term gloss reduction and long-term embrittlement and cracking.

crackling finish: Special effect finish, which is popular above all in the arts- and crafts-market (picture frame coating, restoration of old furniture), because it allows to simulate an antique appearance. This effect is based on a two-layer system: the layer below contains more film former, while the upper layer is filled up with more pigments and fillers (extenders). Because of differential shrinking, in the upper layer fine uniform cracks arise during the drying, which let show through the colour of the layer below.

cratering: One of the most dreaded defects of coating application, because most often it is not retroactively repairable. The cause of cratering is the contact between liquid coatings (but also powder coatings), with higher surface tension, and substances with lower surface tension (→ *polarity*), during the coating application. This may be the effect of an incorrect pre-treatment of the substrate, like insufficient elimination of remains of silicone, grease or hand sweat. In these cases, the coating, due to its higher surface tension, is not able to wet sufficiently such points of the substrate and therefore it withdraws from them. Dell shaped depressions arise in the coating film, which are called "craters". The second occurring possibility is the contamination of an already applied coating by the falling of particles with low surface tension, like silicone, oil, dust, into the still wet film. The third possible occurrence is the presence of incompatible gel particles or an overdose of some additives within the film. The only case when craters are voluntarily generated, is the formulation of → *hammer finish paints.*

critical pigment volume concentration (CPVC): PVC-value, where the amount of film former, which is contained in the coating film, just suffices to fill all hollow spaces between the pigment and extender particles (→ *pigment volume concentration*). By exceeding the CPVC, the permeability of the film increases abruptly and the film properties, which are connected with that, change conspicuously and often negatively. Nevertheless CPVC does not represent a limit never to be trespassed. In the coating technology, there are examples of overcritical pigmented systems, for reasons of price (cheap interior wall paints) as well as of application (putties). Conversely, in other cases, above all when the protection against corrosion is in the most important (→ *barrier effect*), exceeding the CPVC has to be absolutely avoided, with the exception of zinc dust primers.

cross hatch adhesion test: The most frequently used method for determining the adhesive strength of coatings. This means the adhesion of the film to the layer below as well as the cohesion within the film itself. This method is described in the DIN EN ISO 2409 (former DIN 53151). With a one- or a multi-blade knife, a cross hatch pattern is cut on the coating surface. Thus, a lot of little squares are produced, whose sides and corners can flake off differently, depending on the adhesion value. The cutting intervals, and therefore the size of the squares, are fixed by the standards, according to the film thickness. Before evaluation, the cutting area has to be brushed. Sometimes, specifications require an additional step in the form of the detachment of a strip of adhesive tape. In the DIN-standard the damaged patterns are described by values from 0 (0 % detachment) to 5 (→ 65 % detached).

cross-linking: Chemical process (hardening), whereby starting from chainshaped film former molecules, a three-dimensional molecular network is formed. During the cross-linking reaction, thermoplasts are turned into thermosets (→ *thermoplasts and thermosets*) and therefore they lose their solubility and fusibility. This process takes place through all chemically curing coating systems. In this regard, it is not important if the cross-linking is caused by the impact of air oxygen, by stoving or by a two-component reaction.

cross-linking density: It expresses how "small meshed" the molecular network of a coating film is, which has been formed by a cross-linking reaction. The cross-linking density refers to the number of network chain segments per unit volume of the polymer, number which greatly influences the film properties. Generally it is true that the higher the cross-linking density, the higher the chemical resistance and the hardness of the coat. In contrast to this, the coat's elasticity and adherence are mostly reduced (→ *over cross-linking* and *undercured film*).

cross-linking reaction: Chemical reaction, which, during the curing process, leads to a three dimensional molecular network in the coating material (→ *cross-linking*). The cross-linking can be based on one of following reaction types: → *polyaddition*, → *polycondesation* and → *polymerisation*.

cross-linking velocity: It depends strongly on the reactivity of the reactants and on curing conditions like temperature, moisture and presence of catalysts.

curtain coating: Application method, principally used for the coating of wood panels. The coating material is continuously transported by a pump from a supply vessel to a pouring head with an adjustable slit. The slit can be regulated, so that an uniform coating curtain is formed. By means of conveyor belts the work pieces, which are to be coated, are transported at constant velocity through the coating curtain and automatically coated. The thickness of the film depends on the belt speed and on the slit regulation. The excess of coating material, which does not meet the object, flows into a collecting pan and is transported again to the supply vessel. Thus flat, but not necessarily absolutely plane work pieces, can be coated in any film thickness. Beyond the older, unsaturated polyester systems, nowadays unsaturated acrylic resins are employed which, according to their suitability to → *polymerisation*, can be cured in a few second by means of ultraviolet radiation (→ *radiation curing*).

curtaining: Coating defect, whose causes may be found in the coating composition as well as in the application. Generally, curtaining (also sagging, tearing, dripping etc.) appears if during the coating process a too thick wet film is formed, which, beyond levelling, shows a tendency to sagging. This may depend on the too slow evaporation of solvents, during the flash-off phases, for example if the coating material contains too much high boiling solvent or if the spray booth is too cold. But it may also be caused by a wrong spray gun regulation or by an incorrect application technique ("too wet").

cyclised rubber: Derivative of the natural rubber. By means of a special process (cyclisation) the linear molecular shape of the natural rubber is transformed into a honeycomb structure which is almost free of double bonds. The few and still available double bonds can be used , after the physical drying, for oxidative cross linking. Thus, films with outstanding water resistance and chemical stability may be formed (→ *chlorinated rubber*). However, differently from the chlorinated rubber, cyclised rubber shows a heat stability about 130 °C and an absolute physiological harmlessness. Hence it is principally applied in the food industry (for example, coating of dairy plants, pipelines for juices etc.).

D

danger classes: Classification system for hazardous substances according to the safety and product regulations which rule their handling and storage, for example, the Betriebssicherheitsverordnung BtrSichV (Operating Safety Ordinance) and → *GHS*.

danger of explosion: It exists by mixtures of combustible substances, within a determined concentration range (→ *explosion limits*) and air. The combustible substance can be a gas (for example methane), a solvent vapour (for example aceton) or a combustible dust (for instance carbon black). Strictly speaking, an explosion is a combustion reaction which takes place very fast and produces, in an extremely short time, much energy and gaseous reaction products. Together the two cause a shock wave with destroying power.

dangerous freights: Term, which defines dangerous substances and their preparations (for example certain kinds of coating materials), if they are transported. There are different regulations which prescribe the labelling and the container choice for dangerous substances, depending on the route of transport. The most important are: ADR for transport by road, RID for transport by rail, IMDG-code for transport by sea and ICAO-TI for the international aviation.

darkening: Negative property of certain film formers and pigments, which is sometimes connected with insufficient light stability, but occasionally also with after-curing mechanisms. Non-stabilised lead chromates and molybdates darken for example during weather exposition, because the yellow or red compounds of the hexavalent chromium turn incrementally into green compounds of the trivalent chromium. On the other hand, phenolic resins can also darken in the absence of light, above all due to the effect of temperature.

daylight fluorescent paints: Coating materials, which contain → *daylight fluorescent pigments*. Daylight fluorescent paints are predominantly used for decoration purposes (shop windows, scenes etc.) and are mostly applied by means of aerosol cans. However, there are also reserved hues (red and orange), which are exclusively employed as warning colours for fire brigade vehicles and ambulances. In these cases, they are applied by means of spray guns onto a white surfacer, otherwise the hue of the non-hiding fluorescent pigments would be influenced by the colour of the substrate, and afterward they are over coated with a protecting layer of clear coat.

daylight fluorescent pigments: Colorants, which are produced from special dye stuffs and cross-linking amino resins. The transparent dye stuffs, plus light stabilisers, are embedded into the resin, which, after its cross-linking, is ground to a powder. The pigments so produced are obviously susceptible to sharp grinding, for example by means of bead mills, because the pigment particles could be broken down in an undefined way. See → *daylight fluorescent paints*. During irradiation with white light, daylight fluorescent pigments show a luminescence, which produces particularly shiny colours. The reason for this phenomenon is that some wavelength ranges of the white light (for example a blue light), but also ultra violet light, are first absorbed by the dye stuff and emitted again with other wavelengths (for example as a red light). Thus the red light, which in our example is reflected by the pigment, is more intensive than it would be expected by a normal red pigment. This phenomenon, also denominated "daylight fluorescence", is of limited duration, because the exposition to ultra violet rays provokes a degradation of the dye stuff in spite of light stabilisation.

DD-coating: Abbreviation for "Desmodur-Desmophen"-coating. This term traces back to the first isocyanate based enamels, which were developed by the Bayer company. Because of the patent, for years the term had been synonymous with all polyurethane systems. Today it is not used often.

decomposition temperature: Temperature, at which the chemical decomposition begins of → *thermosets*, high molecular, linear polymers and other chemical compounds which cannot evaporate and boil. Above this temperature, cracking processes take place, leading to molecule fission and separation of low molecular components. Obviously, coating materials should not be exposed to working temperatures above the decomposition point of the film former. Organic pigments also show a decomposition temperature, which has to be taken into consideration during their employment. Conversely, for most inorganic pigments and fillers (extenders), the decomposition temperature is so high that they are practically suitable to all kinds of coating application. There are only a few types which, when above relatively low temperatures, tend to a crystalline conversion or to expelling of bonded water (→ *iron oxide pigments*).

decorative paints: Coating materials for the manual application by brush or roller. As film formers they contain mostly oxidative drying → *alkyd resins*, but wood glazes based upon acrylic dispersions also belong to this group.

defective colour vision: Anomaly of the eye, based on a defect of the biological sensors (cones), which are sensitive to coloured light. If, for instance, the red light sensitive cones do not work properly, the red contents of a hue is undervalued and the colour appears greener. This is also true for other colour ranges. Defective colour vision is more common than commonly believed. It is more frequent in men than in women. In the fields of the coating technology which depend on colour accuracy, such as tinting, colour formulation and colour inspection, absolutely perfect colour perception on the part of the collaborators is expected. This can be evaluated by means of special colour vision tests.

defoaming agents: Additives, which facilitate the bursting of bubbles contained in the coatings. Defoaming agents are above all indispensable in water-borne coatings, where a particularly strong tendency to foam formation exists, because of the use of emulsifiers with soap characteristic. The defoaming agents are interface-active substances, whose defoaming effect has a specific influence on the surface tension. Mineral oils are often employed, but silicone and fluorine modified compounds are also important in this regard.

degreasing: Important pre-treatment step, before coating of work pieces. Above all in the case of metallic substrates, the presence of drawing lubricants and conservation oils is to be expected. These substances have to be completely removed before sanding. Grease traces but also silicone residues or sweat on the surface which has to be coated, may significantly worsen the coating adhesion.

degree of fineness: Qualitative criteria for the classification of powder- and platelet-shaped coating raw materials, first of all fillers (extenders) and aluminium bronzes, which during their production are separated into different particle size fractions, by means of screening methods. For example, regarding the extenders the following categories are identified on the basis of their degree of fineness: coarse grinds with particle sizes above 250μm; medium grinds with particle sizes between 50 and 250μm; fine grinds with particle sizes between 10 and 50μm; ultra-fine grinds with particle sizes below 10μm.

degree of grinding: → *grittiness*

demineralised water: Also called "desalinated water" (→ *water, desalinated*). As this expression makes clear, such water is free of dissolved salts and therefore also of the ions, which are derived from the salt dissociation. In the coating technology this condition must be achieved for water-borne coatings, because there ions like calcium and magnesium may negatively influence the stability of the film formers. The demineralisation takes place mostly by means of ion exchangers.

density: Quotient obtained by the division of the mass and the volume of a defined portion of a substance. The density is mostly expressed in g/cm^3 or kg/dm^3. It is an index unique for each substance and is widely used to evaluate the purity of substances or the composition of mixtures in chemical processes. Moreover, density has major importance in process engineering, for example if flow quantities of pipelines, pumps or, in the case of the coating technology, of dispersing units have to be calculated. Since the density depends on the temperature, the required temperature must be strictly maintained during the evaluation.

dew point: The temperature, at which moisture begins to condense in the air. During the coating application, above all in the case of coatings containing highly volatile solvents, permanent cooling of the coating surface takes place. This is caused by heat deprivation due to solvent evaporation, because liquids need a defined amount of energy to change into vapour. If the application is carried out under disadvantageous ambient conditions (low temperature and high air humidity) or if the work piece which has to be coated has not been sufficiently preheated, the additional cooling effect caused by the application may bring the temperature to fall below the dew point. In turn, this brings about the separation of water from the air, which forms a layer on the still wet coating film. Depending on the composition of the coating, this water may cause incompatibility, producing a whitely layer on the surface (→ *blushing*). To avoid such damages, one can add to the coating solvents like butanol, which with water generates a low-boiling azeotrope. Thus the water can be quickly removed from the coating system.

diatomaceous earth: As a natural amorphous silicone dioxide, generated by silicification of microorganisms, diatomaceous earth is primarily used as a cheap flatting agent in dispersion and road line paints.

dibutyl phthalate (DBP): Together with dioctyl phthalate, DBP represents the most important plasticiser among the phthalic acid esters. DBP shows outstanding →*gelling power* and wide compatibility with other coating raw materials. The main fields of application are the plasticising of cellulose nitrate, polyvinyl chloride (PVC) and different copolymers, in addition to the manufacturing of pigment preparations.

dicarboxylic acids: Organic acids, which have two carboxyl groups. The dicarboxylic acids, predominantly the phthalic and the maleic acids, are basic components of the polyesters.

DIN-chromatic-system: The base of this colour system is the DIN-standard 6164, part 1, where the scientific derivation of the chromaticity diagram, for the standard illuminant D 65 and the so called 2° standard observer, is described. This chromaticity diagram, which refers to a tristimulus values system, established in 1931 by the CIE (Commission International de l' Eclairage), was the basis for the development of more modern colour systems, having a better correspondence to the sensitivity of the human eye (→ *CIE L*a*b* system*).

dioctyl phthalate: It shows a gelling power similar to DBP (→ *dibutyl phthalate*). However, it is less volatile than the latter and this characteristic plays an important role in the long term stability of the plasticising. The application fields are comparable to those of DBP.

dipole: In chemistry this term is used for molecules which show a either permanent or an induced electrical dipole moment, where the centre of negative charge distribution does not coincide with the positive one. The consequence is the presence of a negatively and a positively charged pole in the molecule, similar to a magnet. The dipole moment is a scale for the dipole strength and depends on the nature of the atoms involved as well as on the symmetry of the molecules. With increasing dipole strength, the forces of attraction between the opposite dipole charges in the molecules grow (→*adhesion* and →*cohesion*) as well. In the field of coating technology this fact is of major importance for all solution, wetting and adhesion processes.

dipping: One differentiates between two main processes: conventional dipping and →*electrodeposition coating*. Both have in common that the work pieces, which have to be coated, are completely immersed in a coating bath with constant composition and can be entirely coated in a single process step. While in electrodeposition electric current is necessary to carry out the deposition of the coating material, in conventional dipping the film formation takes place through spontaneous adhesion of the coating droplets on the object surface. Conventional dipping can be differentiated between solvent and water-borne systems. Since the combustion resistant so called trilene dipping coatings are hardly in use today, because of their content of trichlorethylene which is harmful to the environment, the water-borne systems are the innovative choice. Among other things, the innovation depends on safety aspects, since the dipping normally takes place at higher temperatures, in order to shorten the drying time. The dipping coating installations are mostly fully automated, closed systems with high operational capacity. The bath composition is permanently controlled and replenished by addition of new coating material and solvents. A permanent coating agitation makes sure that no settling of pigments and fillers (extenders) takes place and the supplements of new material are immediately homogenised. A disadvantage of the conventional dipping is the relatively non-uniform film thickness, which is characterised by sagging and dropping, during the emersion of the object from the dripping bath.

dirt repellent: Property of coatings, which either dispose of self-cleaning mechanisms (→ *nanotechnology*) or contain additives causing an anti-adhesion effect on the coating surface. In this regard, anti-graffiti coatings today are particularly sought, as they contain special slip additives, for example on the basis of polydimethyl siloxane. During drying, these concentrate on the coating surface and smooth it, so that the dirt or the graffiti coating cannot find mechanical bonds to get a firm hold. Moreover, polysiloxane lowers the surface tension of the film and makes such a wetting by external substances more difficult (→ *wetting*).

di(s)azo pigments: Organic pigments, which contain in their molecule two azo groups. They differ from the ordinary azo pigments (also called monoazo pigments) above all because of better solvent resistance (→ *bleeding*).

dispersing units: devices, which are used to break down the agglomerates of pigments and fillers (extenders) in coating materials. The breaking down succeeds by the application of mechanical and thermal energy. Concerning the different dispersing units, see → *dissolver*, → *triple roll mill*, → *ball mill* and → *bead mill*.

dispersion: A generic term to express different distribution states of film formers or pigments/fillers in liquids, mostly water, which cannot be individually discussed in this compendium. In the colloquial language of the coating technology, the term "dispersion" refers in general to polymer dispersions (also called "primary dispersions"). That means almost spherical polymer particles, which are dispersed in water and produced by the emulsion polymerisation method. Such dispersions are mostly employed in dispersion paints and synthetic resin plasters.

dispersion paints: Physically drying systems, which contain aqueous polymer dispersions as film formers. According to the required properties, vinylacetate copolymers or acryl/styrene and styrene/butadiene copolymers are predominantly used. Dispersion paints are extensively filled with pigments and fillers (extenders) and thus they show high pigment volume concentration values (PVC). As a rule in interior wall paints the CPVC (→ *critical pigment volume concentration*) is even exceeded. In order to maintain in floatation the big amounts of pigments and fillers (extenders) and to achieve a better workability, dispersion paints are equipped with thickeners or thixotropic agents. The film forming, after the water evaporation, is supported by the addition of → *coalescence agents*. Other additives employed are wetting agents, defoamers and preservatives. The main application fields of dispersion paints are interior and exterior wall paints as well as synthetic resin plasters.

dispersion process: Operation breaking down the agglomerates of pigments and fillers (extenders) in coating materials, to obtain as far as possible only primary particles and eventually fine aggregates. A main precondition for a successful dispersion process is a sufficient wetting of the pigment particles by means of the binder/solvent-mixture (→ *wetting*).

dissolver: Also called agitator. It is the most economical among the dispersing devices, which are commonly used in the coating technology. Its advantages include first of all relatively low energy and time expense and simple handling. It consists of a shaft, with continuously adjustable speed, which can be fitted with toothed discs of different diameters. The shaft can be automatically lifted or sunk in a dispersion vessel, containing a mill base. The dispersing effect depends on the proportion of the diameters of the vessel relative to the employed disc, as well as on the peripheral velocity of the disc. The viscosity of the mill base and the temperature are also relevant. Since the dispersion effect of the dissolver is exclusively produced by the shearing stress, caused by the flowing liquid layers (→ *bead mills*), a satisfying dispersion result can only be achieved by employing inorganic pigments and fillers (extenders) which are easy to disperse. In the coating technology the dissolver is steadily used for the pre-mixing and pre-dispersion of mill bases. Moreover, it is the main dispersing unit for manufacturing of dispersion and road line paints as well as of most kinds of primers, surfacers and putties.

dissolving power: Faculty of a solvent to dissolve film formers. The dissolving power has to be individually referred to each single film former, because there is no all-round solvent capable of dissolving every material. Of importance for the dissolving power is the relative position of the solubility parameters of the solvents and of the binders to each other (→ *solvent, real*). In essence, the more the chemical structures of the solvent and the film former are similar to each other, the greater the success of the dissolving process. This success depends on the ability of the substances in question to create interactions between their molecules.

doctor blade: Device for the application of coating layers with defined wet film thickness. There are different designs of doctor blades, either denominated according to their shape (frame, wedge or spiral doctor blade) or to their function (flow and running doctor blade). Most frequently, frame doctor blades are used, which consist of a stainless steel frame whose edges show different gaps with defined heights in μm. So it is possible to apply with the same doctor blade films of different wet film thickness. Normally the application takes place in a manual fashion. Filled with coating material, the frame doctor blade is pulled as uniformly as possible in the opposite direction to the gap with the required height, across a metal-, a glass-panel or a paper board. There are also motor equipped doctor blades, which are quite suitable for application on paper board, if a particularly uniform film thickness is required, for example in the case of comparing tests.

dolomite: → *calcium carbonate*

domestic appliance coatings: Often also called "coatings for white goods", they have to fulfil high standards regarding general wear resistance and, most importantly, resistance against food and all kinds of detergents. In the case of washing machines and dish washers, sufficient resistance against hot alkaline substances is required. Depending on the fields of application, often coil coating materials (→ *coil coating*) and powder coatings are employed. For the specialized application of interior coating of cookers, high heat-resistant enamels based on silicone resins are preferably used.

double bond: Chemical term, which describes the connection of two atoms by means of two common electron pairs. Double bonds are particularly important in organic chemistry, especially between two carbon atoms, because they form the basis of all polymerisation processes.

driers: They serve to accelerate the oxidative cross-linking of drying oils and alkyd resins (→ *oils* and *alkyd resins*), which without them would take a very long time. Chemically, driers are soaps, that is metal salts of long-chain carboxylic acids, whose drying effect depends only on the type of metal. The acid moiety provides only the drier molecules with the necessary solubility and compatibility in their application system. Generally, driers are divided into primary and secondary (or auxiliary) ones. The primary driers are actually able to catalyse the cross-linking reaction. The precondition for obtaining this effect is the presence of a metal, which can change between two valences. As a result of the reaction with oxygen, hydroperoxides are temporarily formed in the molecules of the drying fatty acids, whose decomposition supplies the radicals necessary for the cross-linking (→ *radical*). The valence change of the metal in the drier accelerates, by cession or assimilation of electrons, the formation of radicals and therefore of the drying process. The most important metals of primary driers are cobalt, zirconium and manganese. Lead, which in the past was often employed due to its good through drying, is now not used any more because of its toxicity. The secondary driers contain metals, which show only one valence and therefore are not able by themselves to influence the drying velocity. However, in combination with primary driers, they can improve the drying efficiency. This result is mainly caused by their function as wetting agents, which insures that the primary driers in the long run are not absorbed by the pigment particles and do not loose their effect. The most important secondary driers are calcium and zinc octoate and naphthenate.

driers, dosing of: Addition of the suitable amount of driers in oxidative drying systems, that is in oil varnishes and alkyd resin coatings based upon drying fatty acids (→ *oxidative drying*). The right dosing of the driers has outstanding importance for these systems. A wrong dosing can lead to serious coating damages like for example → *wrinkling* or a considerable drying delay.

dripping resins: Essentially they fulfil the same function as the → *impregnating varnishes*, but they are used to stabilise electrical coils which are subject to particularly high mechanical exposures (for example rotors of electric motors). They are normally based upon unsaturated polyesters with → *reactive thinners* and are let drip in the form of a thin jet into the rotating electrical coil, until this is completely filled. From this process is derived the term “dripping resin”. The curing then takes subsequently place by means of UV-radiation or of resistance heating caused by current passing through the workpiece.

dry film thickness: →*film thickness measurement*

drying and hardening methods: They differ from each other in that in the drying methods the film-forming is caused first by evaporation of solvents, including water, (→*drying, physical*) and eventual cross-linking processes take place simultaneously or subsequently to that. In the hardening the film-forming is produced practically only by chemical cross-linking. A common classification into categories is: drying by heat sources (→*convection drier*), drying by radiation (→*infrared drying*) and hardening by radiation (→*electron beam curing* and *UV-curing*). Electrical drying processes, like resistance drying and inductive drying, play only a subordinate role and therefore they should not be described in detail in this compendium.

drying, chemical: Usual but improper term for "chemical cross-linking". It describes the process of progressive solidification of films in chemically hardening coating systems, as a result of increasing cross-linking among the binder molecules, until the final stage of a three-dimensional network is reached. The main part of chemical cross-linking takes place after the conclusion of the physical drying.

drying, forced: →*forced drying*

drying oils: →*oils*

drying, oxidative: →*oxidative drying*

drying, physical: The phase of film formation, which is based only on the evaporation of the solvents. In case of chemically hardening systems, it happens before and/or in parallel to chemical cross-linking (→*drying, chemical*). In the case of physical drying, the distinction is usually made among different stages, which are dependent on the progression of solvent evaporation. A common classification in the application process is: initial drying,→*dust-free drying* and *trough drying*: Film formers, which dry only physically, can be dissolved again by means of suitable solvents. Examples of physically drying systems include: dispersion paints, nitrocellulose combination lacquers, metallic base coats and polyvinyl chloride copolymers.

drying time measurement: There are different methods for monitoring the drying process. Accurate determination of the different drying stages is made possible by a drying time recorder. With a special doctor blade, a coating film with defined film thickness is applied onto a narrow glass bar. The glass bar is placed in the drying time recorder, so that by means of a lever a pointed mechanism, which is fixed on a movable sledge, can be let down exactly onto the applied coating film.

Then the recorder is started. At constant speed and for a determined period of time, the recorder draws the sledge with the point along the drying film. According to the scratches left behind by the point, the different drying stages can be reconstructed. Another and simpler method is to determine the degree of dryness, according to DIN 53 150, where the "stickiness" of the coating surface represents the measure for the evaluation of dryness. Up to degree of dryness 2, fine glass beads are scattered onto the coating film. One evaluates how easily they can be removed from the film by the tilting of the test panel or by means of a brush. Beyond degree of dryness 2, paper is printed on the film using a test stamp. The evaluation regards how many paper fibres are left behind on the coating surface. Other methods, for example the "rolling ball" according to Rossmann, are seldom used. However, irrespective of the test method, it should be considered that the film-forming process is strongly influenced by the film thickness and by the existing temperature and moisture conditions. In order to get reproducible results it is recommended that the tests be carried out in a conditioned room.

dry sanding: Pre-treatment method for wood and metal substrates (→ *wet sanding*). In regard to wood wet sanding is not advisable due to the absorbency of the substrate and, in case of a metal, particularly iron and steel, the contact with water could cause the formation of a rust layer. Abrasive papers with corundum grains are principally used (→ *sandpaper*) in dry sanding. Paper for wet sanding with silicone carbide grains may also be employed for dry sanding.

dry spraying: Special spray technique, which is used during the application of effect base coats to influence the metallic or pearlescent effect. It is based upon the evidence that a metallic colour looks the "more metallic" (higher mirror effect) the finer the coating is atomised during its application. The solvent of smaller coating droplets evaporates quicker on the way to the object to be coated. So the aluminium platelets are promptly fixed in a horizontal position in the drying coating layer. Conversely, aluminium platelets in a wet film, during the drying process, are destabilised in their horizontal position by the whirls caused by the evaporating solvents. This leads to a worsening of the mirror effect.

dust explosion: Explosion, caused by combustible dusts which are dispersed in the air. To start the combustion process, which leads to the explosion, an ignition source is necessary. This can be a flame, but also a spark, originated by electrostatic charge. In the coating industry, the danger of dust explosion is only created by the processing of powder coatings. Therefore, in this case appropriate safety measures have to be taken (→ *electrostatic charge*).

dust-free: Important precondition for achieving satisfactory surfaces by coating application. Depending on the exigencies the related technological expense may be very high. The maximum complexity is found in the automotive OEM coating and in the automotive refinishing. The application areas of the modern automotive coating processes are practically "clinically clean". People can enter the areas only dressed in overalls of dust-free material and after passing through ad hoc air channels.

dust-free drying: First phase of the physical drying of a coating material (→ *drying, physical*), at the end of which the film has become solid enough that falling dust cannot stick any more. Quick dust-free drying is important for coating materials which are applied under improvised, mostly manual conditions, because it determines the dirt susceptibility of the wet paint.

dust mask: → *breathing protection*

dye stuffs: Colorants, which are soluble in the application medium. In the coating technology dye stuffs play a subordinate role, because as a result of their extreme fineness in the state of solution, they do not show sufficient light and weather resistance in most of the coating application fields.

E

earthing: Leading off electric charges from an object by means of a conductor, which is connected to the earth. In the coating technology, earthing measures can be found very often. On the one hand they serve for safety, that is all machines, dispersing devices and pipelines have to be earthed in order to avoid the danger of fire, caused by electrostatic charge. On the other hand, earthing is used in electrostatic application methods, to generate a sufficient potential difference between the object to be coated and the application device (→ *electrostatic application*).

eco friendly: Requirement for all coating materials and systems and their corresponding application technologies, which has recently come to the fore. For several decades, the formulation of eco friendly systems has been the prompter for coating research and development. Above all concerning the → *immission protection*, innovative solutions like → *water-borne* and *powder coatings*, *high solids* and radiation curing systems (→ *radiation curing*) have been developed. But it has also prompted the substitution of single raw materials or raw material groups, which scientific studies have shown to be dangerous to human beings and the environment. Examples of banned substances are: chlorinated and fluorinated hydrocarbons (ozone hole), zinc chromate, cadmium pigments and several lead compounds (toxic or carcinogenic and ground water polluting) and some glycol derivatives (teratogenic).

edge covering: Weak point of many application methods. The conventional spraying processes (high pressure, low pressure and airless spraying) often show the risk of negative edge effects, which are particularly marked in the case of water borne coatings, due to their high surface tension. Thus, during the film forming, the coating inclines to contraction and therefore it retires from the edge. As a result, only a thin layer remains there. The consequence of that is a reduction in the mechanical and corrosion protection. The attempt to balance the edge effects by application of thicker films, leads often to curtaining and sagging. Of course, a higher coating viscosity lowers the danger of edge effects, since it permits the formation of thicker film layers, but on the other hand this can lead to a worse levelling. In the electrostatic application processes (→ *electrostatic application* and *EPS-coating*), at the edges there is a rather clear risk of excess coating. In the electrical field, at sharp edges a particularly high field strength (high concentration of field lines) develops, so that here, as a rule, a higher deposit of coating droplets is expected, in comparison to the remaining work piece surface. The most uniform edge covering can be performed by → *electro deposition coating*. This is one of the principal reasons why, in the automotive coating (OEM) world wide, the corrosion protection primers today are applied exclusively using this method.

effect finishes: Coating materials, whose appearance does not only depend on the usual mechanisms of colour generation, that is light absorption and scattering. In the first place, there is the case of the coatings which are equipped with effect pigments. But in general, coatings which show special surface structures like → *structured paints, hammer finish paints, wrinkle finishes* and *crackling lacquers* also belong to this group.

effect pigments: Common term for all pigments, whose colour is not uniquely produced by simple light absorption and/or scattering. In the first place these are the aluminium bronzes (specular reflection) and the pearlescent pigments (colour generation by light interference, eventually in combination with light absorption). Also newer variants like liquid crystals, aluminium platelets, coated with iron oxide or organic pigments, and daylight fluorescent pigments belong to this pigment group. Innovation in this field is very high, because of the sustained demand for new stylish effect colours.

elaeostearic acid: Important drying fatty acid, which is contained in the wood oil. Due to the presence of three conjugated double bonds, it shows outstanding oxidative drying properties and it tends less to yellowing (→ *linseed oil*) than the linolenic acid. Its main field of use is the production of drying alkyd resins.

elastomers: Polymeric substances, which consist of loosely cross-linked molecules. Under the influence of a tensile stress, the few cross connections can be temporarily stretched, so that the molecule chains are shifted against each other. As soon as the mechanical stress ceases they get back their original shape. Because of the elastic properties these materials are often used in technological applications (sealing washers, cushioning, sound proofing).

electrical conductivity: Important property of coatings, which have to be applied by → *electrostatic application* or *electrodeposition*. Whilst in the electrodeposition process the electrical conductivity is generated chemically, in the case of the electrostatic application the suitability of the coating material to electrostatic charge is a basic precondition. The tendency of a coating material to accept electrical charges depends on its conductivity. If the conductivity is too low, the number of electric charges, which can be accepted from the coating material, is not sufficient to provide to insure the transport of the coating droplets along the electrical field lines. If the conductivity is too high, the possibility of a short-circuit in the conductive coating mist may lead to a breakdown of the electrical field.

electrical insulation coatings: Coatings, which, in the electrical engineering, have the duty to electrically insulate conductive parts from their environment and also to mechanically protect them. In addition to outstanding insulating properties, they have to show a permanent elasticity, good adhesion and abrasion resistance, even after long periods of use (aging). Examples are: wire enamels, used to coat copper wires, and dripping agents, used as casting compounds to reinforce the wire coils of electric motors. As film formers what are most often employed are phenolic resins, polyvinyl formale, polyamide resins, blocked isocyanates and unsaturated polyesters.

electrically conducting varnishes: They have the function to provide a nonconductive substrate with an electrically conducting layer. This need depends on different technological requirements. A good example is the application of conductive primers for plastics. They contain conductive pigments, like carbon black or micas, coated with a tin/ antimony oxide layer, and are used to make possible the electrostatic coating of non-conductive plastic work pieces.

electrochemical series: Classification system, mainly used for metals, which ranges the elements or the ions, containing metal atoms, according to their oxidation or reduction behaviour. The zero point of the series is by definition the hydrogen, which by convention has been assigned (arbitrarily) the potential value zero. A galvanic cell, consisting of a bar of the correspondent metal immersed in its metal ion solution, is connected to a hydrogen standard electrode by means of a conducting wire. One observes in which direction the electrons flow. Additionally, the potential value in volt, which arises in this galvanic element, is measured. The measured potential value is registered as normal potential in the electrochemical series. If the electrons flow from the metal to the hydrogen, the potential value obtains a minus sign and is placed left of the hydrogen. If the electrons flow in the opposite direction, the sign of the potential is a plus and its value is registered right of the hydrogen. The more negative is the normal potential, the "less noble" is the metal, that is, the easier it is to be oxidised. Noble metals like gold or platinum, which stay in the nature unaltered, because they do not show any tendency to oxidation, have always positive potentials. Conversely, metals such as iron or aluminium, which are inclined to oxidation, show negative potential values. In the coating technology, the knowledge of the electrochemical series gives important indications about the oxidation behaviour of the metal surfaces which have to be coated.

electrodeposition coating: Very eco friendly and rational industrial coating process, which is based upon the principle of electrolysis. In an electrolysis bath, ions, that is electrically charged atoms or atom groups, move to an electrode, which is dipped into the bath solution and shows the opposed charge to them. In the electrodeposition coating process, the coating droplets are the charged particles and the object to be coated is the electrode. Due to constant agitation of the coating bath, the coating particles are led near the object with the opposed charge. Depending on the process variant, neutralisation takes place by means of H^+ or OH^- ions, which are produced by the electrolysis of water. Thus, the coating particles lose their → *water solubility* and coagulate on the object surface. The film, which results is very uniform and characterised by extremely good edge covering. The main advantages of this process are low emission and absence of overspray. A disadvantage of the electrodeposition coating is the limited relatively low maximum film thickness, which can be obtained. This is the reason why, to date, with this process a single coating layer can be applied at the most. With the increase in the film thickness, the insulation strength of the film rises as well, until the current flow breaks down.

electrodeposition coating, anodic: In the case of this variant, the coating droplets are negatively charged and the work piece to be coated becomes the positively charged electrode, that is the anode. Stainless steel cathodes are used as opposed electrodes. The anodic electrodeposition coating is the older process and, since it shows weaknesses relating to corrosion protection and throwing power, today it is only employed for the applications where the weather exposition is not a key element (heaters, agricultural implements).

electrodeposition coating, cathodic: Is the newer and more expensive process, clearly superior to the anodic electrodeposition coating in terms of corrosion protection. In the case of the cathodic method, the coating droplets are positively charged and the work piece to be coated is made the negatively charged electrode (cathode). Also, in this process, the opposed electrodes (anodes) are of stainless steel. The superiority of the cathodic electrodeposition coating relative to corrosion protection is the reason why it is used as the main process for the automotive OEM priming.

electron beam curing (EBC): Curing method for coatings, which are suitable to polymerisation, where radioactive beams (β-radiation) are employed as initiators to trigger the radical polymerisation. According to this method, the double bonds of the film formers are divided into radicals, which are subsequently connected with each other to form new simple bonds. This cross-linking process takes place in a split second and, therefore, it is faster than by UV-curing. As a result of the high radiation energy, the EBC-process also allows for a good through curing of pigmented coatings (→ *UV-curing*). On the other hand, the electron radiation can cause the formation of oxygen radicals, which would generate a sticky surface. Therefore, the curing process must be carried out in the presence of helium as a protective gas. In spite of the higher curing speed, the EBC is less wide-spread than the UV-curing, because the screening of the radioactive beams as well as the creation of an efficient helium atmosphere for the coating of three-dimensional work pieces is very expensive.

electrostatic application: Application method, where the strength of an electric field is used for the coating atomisation and the transport of coating particles. For the electrostatic application devices are employed, which have sharp edges or cone points with a high concentration of field lines. Thus the field strength is highly increased at these special spots. According to the different processes used, one works with voltages between 80 and 160 kV but with very low current intensities, in the range from micro- to milliampere. The work piece to be coated is grounded and represents therefore the positive pole, whilst the application device is charged negatively. The electrostatic application is suitable for liquid coating materials as well as for powder coatings (→ *electrostatic powder coating*). The atomisation of the coating and the transport of the coating droplets take place by the use of kinetic energy, which, according to the application variants, is employed in different ways. For example, in the case of the rapid-rotation bell, the centrifugal force comes into operation. Concerning the so called "electrostatic spraying", the atomisation principles of → *high pressure spraying* or of → *airless spraying* are most relevant. The electrostatic charge of the particles in the rapid-rotation bell takes place at the sharp edge of the bell. In the high pressure and airless spraying, internal electrodes or needle-shaped electrodes near the nozzle provide the charge. The coating droplets move along the field lines to the work piece and deposit there as a film. After that, the physical drying of the film (for example in case of metallic base coats) or its cross-linking by stoving is carried out. Specific advantages of this application method are the very low overspray (little material loss) and that, due to the so called throwing power, the back of the work piece is coated as well. A disadvantage are the problems associated with the coating of voids (→ *Faraday cage*).

electrostatic charge: Physical phenomenon, which happens as a consequence of friction and separation processes between substances with low electrical conductivity. Such processes provoke an interchange of electrical charges, which cannot be transported within the non-conductive substance, that is the insulator, and therefore increasingly accumulate. If the charged insulator approaches a conductor, a sudden discharge with spark formation may take place. The sparks are able to set on fire combustible mixtures of solvent vapours and air or to cause explosions. In the coating industry, a lot of stringent measures are prescribed to prevent electrostatic charge, including conductive footwear, earthing of machines, devices and pipelines. Conversely, in the case of the electrostatic application processes, the electrostatic charge is produced artificially, in order to transport the coating droplets by means of an electrical field (→ *electrostatic application* and *electrostatic powder coating*).

embrittlement: Result of → *overcuring* or of a degradation process of coating films, produced by the weather. Sometimes a plasticiser migration may also be the cause of embrittlement. Visible marks of embrittlement are cracking, loss of adhesion and flaking.

emission: The harmful substances, in the form of vapours, dusts, noise and radioactivity, which are emitted by a plant and reach the atmosphere.

emulsifying agent: Additive, to produce systems within which a liquid is very finely distributed in an other one (→ *emulsion*). The molecule of the emulsifying agent contains groups which have a strong affinity to the dispersed droplets, while others molecule moieties are very compatible with the surrounding liquid. Thus, the emulsified droplets are thoroughly wetted and can be dispersed in a stable way.

emulsion: System, consisting of two liquids, one of which is very finely dispersed in the other (generally water). Since, as a rule, the diameter of the droplets is bigger than 0.1 μm, they are able to reflect the light. Therefore the appearance of emulsions is cloudy to milky and the milk itself is a good example of a natural emulsion. The stability of the emulsion is guaranteed by the layer of the emulsifying agent, being deposited on the droplets surface, and/or by electrostatic charges. If this layer is removed or damaged, for instance by too high temperature or by the effect of solvents, the droplets cannot swim in the other liquid any longer and, therefore, coagulate ("breaking" of emulsions). In the coating technology, supplementary additives, such as protective colloids, are employed to stabilise emulsions.

environmental protection: Important purpose for all phases concerning the development, production and application of coating materials. In addition to its economical and technological aspects, environmental protection today plays a key role in the coating technology, as coating manufacturers have to comply with many different national and international regulations, to obtain the authorisation to use their coating products.

epoxy resin esters: Epoxy resins which have been modified by fatty acids, similar to the alkyd resins. Depending on the kind and amount of the employed fatty acids, they may be suitable for the oxidative drying or the stoving cross-linking (→ *alkyd resins*). But when compared with the alkyd resin types, the epoxy resin esters show a better profile, that is they are harder, more elastic and have better adhesion and corrosion protection properties. On the other hand, disadvantages include their tendency to chalking and their higher price. Oxidative drying epoxy resin esters are mostly used for primers and surfacers (fillers). The main application of the stoving types is the creation of highly elastic and well adhering tube enamels, deep-draw and punchable finishes.

epoxy resin powders: Historically, they are the oldest powder coating systems. They are generally subdivided into pure epoxy powders and hybrid powders. The pure epoxy powders are characterised by high wear resistance and good utility properties, though they are not suitable for outdoor exposure (chalking). They are cross-linked with dicyandiamide or polyanhydrides. In the case of the hybrid powders, a second film former component, an acid polyester, is used as a hardener for the epoxy resin component.Weather resistance improves with the presence of the polyester but without achieving the level of the pure polyester powders.

epoxy resins: Group of film formers, which are characterised by high chemical resistance, excellent adhesion and outstanding corrosion protection properties. A problem with epoxy resins is their tendency to → *chalking*, if they are exposed to weathering as coating materials containing pigments. They consist of linear molecules, which show at both ends two oxirane rings (epoxy groups). Since these may provoke an addition reaction, at room temperature with suitable partners (for example polyamines or polyamides), epoxy resins can be used for the formulations of two-pack systems. On the other hand, they may also be cross-linked by means of their OH-groups in combination with other film formers, at higher temperatures as stoving systems. The main fields of application of the two-pack systems are the heavy duty corrosion protection (→ *corrosion protection, heavy duty*) and the building protection. As stoving systems, the epoxy resins are used primarily in combination with phenolic resins, melamine resins and blocked isocyanates, for can coating and coatings with high chemical resistance. Amine modified epoxy resins play a decisive role as film formers in cathodic electrodeposition (→ *electrodeposition coating, cathodic*).

EPS-coating: EPS stands for "electrostatic powder spraying" and it is one of the most important application methods for powder coatings. One distinguishes between the corona and the tribo powder spraying. In both cases the work piece to be coated is earthed and therefore it is made the counter-pole to the charged powder particles. In the corona process, the powder is negatively charged by way of a high-voltage generator (voltage: 90 to 120kV). In the tribo electric process, the charge takes place by friction of the powder against the inside of a special plastic spray gun, whereby the powder loses electrons and therefore charges itself positively. The powder transport to the spraying element in both cases is carried out by means of compressed air. The powder particles move from the spraying element to the work piece along the electrical field lines, deposit onto the object and adhere to it through electrostatic forces of attraction. Afterwards, the melting of the powder at the object surface (levelling) and its cross-linking are carried out simultaneously by stoving.

Erichsen indentation: Method to determine the deformability of coating films. The coated test panel is clamped in a holding frame. A round-ended steel indenter is pressed with constant velocity into the uncoated back of the panel, whereby on the panel surface a protuberance arises and the coating layer is increasingly deformed, till it cracks. Using a magnifying-ocular, the indentation (in mm) is measured as soon as the first cracks appear on the surface of the coating.

esterification: Chemical reaction between an acid and an alcohol, which leads to the formation of an ester, with simultaneous separation of water (→ *esters* and *polyester resins*).

esters: Organic compounds, which are generated by the reaction between alcohols and acids with separation of water. Simple esters (monoesters) play an important role as solvents. Above all is butyl acetate, which among other things is the main substitution product for xylene (→ *aromatic free*), and ranks among the most important medium boilers. Macromolecules, which contain several ester groups, represent one of the most important binder groups → *polyester resins*.

etherification: Chemical reaction between two alcohols, which leads to the formation of an ether, with simultaneous separation of water. The etherification, which takes place among the OH-groups of binder molecules, is the basis for cross-linking in most stoving systems.

evaporate: Progressive transformation of a liquid into a gas. Simple evaporation should not be confounded with boiling, which physically means the conversion of a liquid into a vapour at its temperature of ebullition. Different from boiling, the simple evaporation (volatilisation) takes place only at the surface of the liquid. It is the stronger the larger the surface area is. With increasing external pressure, the tendency of the liquid to evaporate decreases. For the coating technology, the evaporation behaviour of solvents is of major importance relative to the application properties of coating materials (→ *evaporation index*).

evaporation index: Coefficient for evaluating the volatility of a solvent. It is determined according to DIN 53170 and expresses how many times more slowly a solvent evaporates than diethyl ether. The test substance is dropped onto a filter paper and the time is recorded, of how long it takes for the wet solvent spot at the paper surface to completely disappear. The same procedure is carried out with diethyl ether. The evaporation index has great practical importance, since it permits the comparison of the evaporation behaviour of different solvents at a glance, especially because the values of the most important types are registered in tabular form. However, due to the different →*solvent retention* in the drying coating film, the evaporation index does not permit a direct prediction of the real evaporation velocity.

exhaust system: Technical device to reduce the concentration of harmful vapours, dusts and/or aerosols, like spraying mist, in the air of a workroom. In the coating technology, exhaust systems are for the protection of human beings (danger of inhalation of solvents and pigment dust, →*MWC-value*) as well as for the prevention of fire and explosion, due to the too high solvent concentration. Finally, they guarantee better coating results, since they avoid the pollution of wet coating layers by coating mist and dust.

explosion limits: They refer to the concentration range of a combustible substance in the air, where an explosion may take place. There is a lower and an upper limit to the concentration. If the concentration is under the lower limit, the amount of combustible substance in the mixture with air is too little to cause an explosion. If the upper limit in the mixture is exceeded, there is too little oxygen to provoke the explosion reaction.

extenders: Term, used in the coating technology for low-priced substances, which are primarily employed to cheapen coating materials. Depending on the formulation, these can be solvents or fillers. In the case of solvents, as a rule they represent →*non-solvent* components for thinners, which can only be added to the coating materials in such amounts that no separation or flocculation of the binder or other coating components may take place. The solvent extenders have to evaporate faster than the real solvents (→*incompatibility*), during the film forming. Low-priced fillers are used above all for industrial coatings and dispersion paints as extenders to reduce the amount of white pigments. In this context, especially in heavily-filled internal wall paints, titanium dioxide is often partially substituted by calcium carbonate. Finally, it is important to mention that in the colloquial coating language the term "extender" or "pigment extender" is often incorrectly used instead of the right term "filler", which expresses the basic purpose of application for this group of coating components (→*fillers*).

extruder: Technical device for the production of powder coatings and plastics. It consists of a heatable cylinder, in which one or two screws rotate so that, similar to a meant mincer, the contents are kneaded and transported to the cylinder exit. The solid components of the powder coating, that is binder, pigments and fillers (extenders), hardener and eventually additives are premixed and then put into the extruder. During the kneading, the binder melts and wets the pigments and extenders. Through the kneading rotation of the screws, in addition to the transport of the material an intensive dispersion of the pigments takes place as well. At the extruder exit, the hot paste-like material is pressed through a nozzle, than led to a cooling roll, where it becomes solid again. This procedure can be repeated several times, until the required degree of dispersion is reached. In practice, this happens rather seldom

F

Façade paints: → *wall paints, exterior*

fading: Change of a colour to a "paler" hue because of light exposure. Indirectly, fading gives information about the light fastness of the pigments employed. However fading depends also on the depth of the colour involved. → *Chalking*, during weather exposure, can intensify fading.

Faraday cages: Hollow spaces (voids), where the lines of an electric field pass by without entering them. Faraday cages represent problematical zones for the various electrostatic application processes. Often, the interior coating of such voids is possible only either by means of auxiliary electrodes or by the support given by kinetic energy of the coating particles generated by other methods.

fatty acids: They consist of mostly linear molecules, which contain respectively only one acid group (carboxyl group). The number of their carbon atoms is 12 or more, while the types which are interesting for the coating area normally contain 18. In addition to simple bonds, they can have one or more double bonds between their carbon atoms, which influence decisively their drying properties, that is the capability of cross-linking by the aid of oxygen. The name "fatty acids" is due to the fact that they are components of natural fats and oils, from which they can be extracted by saponification. In the coating industry the fatty acids still play an important role for the production of fatty acid modified polyesters (→ *alkyd resins*).

fatty acids, saturated: Types, which contain no double bonds and therefore are not able to oxidative drying. Natural, saturated or low unsaturated fatty acids mixtures, such as those contained in peanuts oil or coconut oil, but also synthetic fatty acids which are employed to manufacture alkyd resins for stoving systems.

fatty acids, unsaturated: They contain at least one double bond in their molecule. But they achieve good drying properties only from two double bonds onwards. The most important types for the coating technology are the linoleic acid with two double bonds and the linolenic and the elaeostearic acids with three. They are mainly used for the production of drying alkyd resins for house and decorative paints.

fillers: This term is also used to refer to intermediate coating materials, employed to level out little substrate irregularities. But in this application, the term →*primer surfacers* is more common and should be generally used. The main application of the term "fillers" is reserved for mostly white or greyish coloured, solid components of coating materials (→ *extenders*) which, due to their too low refractive index, do not have pigment character and therefore, as a rule, are used in combination with pigments. As the name indicates their main duty is to give the coating material filling power, that is the capability to fill in the irregularities of the substrate. This is possible because of the filler particles, which form a solid structure in the wet coating material, capable to bridge such little irregularities. Moreover, the fillers influence the condition of films in many ways. In heavily-filled systems, they can partly substitute white pigments and therefore lead to a cost reduction; they can influence properties like gloss, grindability, adhesion, and elasticity; in primers, above all platelet shaped fillers contribute to improve corrosion protection (→*barrier effect*).

filling: Last stage in the process chain of coating manufacturing. Before the true filling process, that is the weighing of exact coating amounts in suitable containers, what should also take place is the elimination of impurities in the coating material, by means of different filtration methods (→*filtration*).

filling primers: They combine the properties of a corrosion protection primer with those of a primer surfacer (filler). They are increasingly used in the automotive refinish but also in some industrial coating systems. They are equipped with effective anticorrosive pigments (→*primers*) and at the same time show a high filling power, due to an optimised extender packing. The big advantage of using filling primers is the saving of one complete working operation as well as of the related material costs.

film: The layer, which remains behind on the coated object at the end of the drying or hardening process of a coating material. Practically, the film represents the combined result of the coating process and has to meet all properties required of the coating systems (→ *film properties*), as completely as possible.

film formers: → *binders*

film forming: → *film*

film properties: The film properties of coating materials mainly focus upon the embellishment of objects (colour, gloss, surface structure) and the protection against corrosion, weathering and all kinds of mechanical stress and chemical exposure. In individual cases, other particular properties may be added like electrical conductivity or insulating properties, fire resistance, suitability to deep drawing, etc.

film thickness measurement: One differentiates between dry and → *wet film thickness*. Measurement methods are numerous. The ones to evaluate the dry film thickness are subdivided into non-destructive and destructive methods. Following, there is first the description of the most important non-destructive methods. For ferromagnetic substrates the measurement takes place, according to DIN 50 981, either by using the magnetic adhesion of a permanent magnet or the magnetic induction flux of an electromagnet, both methods depending on the thickness of the non-magnetic coating film. In the case of non-ferrous metals, the film-dependent eddy current loss is measured, according to DIN 50 984. On plastic surfaces, ultrasonic wave methods are increasingly used. The more frequently employed destructive measuring methods are the IG-meter and the light section microscope. The IG-meter is a linearly arranged tripod with fixed outside feet, while the central one is used as movable probe. The dial has a scale with a revolving pointer. To take the measurement, at one point of the film the coating has to be carefully removed to reveal the substrate. Doing this, one has to pay attention not to damage the substrate, because this would alter the measurement. The tripod is set up such that the outside feet stands on the coating surface, while the probe touches the substrate at the laid bare point. The height difference can be read directly on the dial. The light section microscope renders possible the non-destructive measurement of clear coats. However, in the case of pigmented films it is necessary to remove a coating strip to reveal the substrate. The measurement principle is based upon the comparison of two rays of light, being produced by the reflection at the upper and lower margins of the coating film, using a micrometer eyepiece.

filtration: Important process stage of the coating production, which especially serves to purify the coating material before its filling. However, filtration steps can also be anticipated, for example to clarify binder solutions, semifinished products, solvents etc. One distinguishes here between two different main methods: surface filtration (often also called "screening") and depth filtration. In the first case, screens of metal or of synthetic fibres are employed, which have a known mesh size and therefore are capable to separate only particles, which are coarser than the mesh size, in a very selective way. In the second case, thick layers of porous filtration materials are used, which do not have a defined mesh size, so that their retention rate has to be determined experimentally. Concerning the filtration devices, there are also two options: bag and candle filtration. Which of these processes is suitable to which coating material depends on the respective quality requirements.

fineness of grind: → *grittiness*

fire prevention paints: Also called intumescent paints. Paints, which, in the case of fire, can generate a solid carbon foam with high heat insulation properties. This heat barrier retards the moment when the substrate to be protected either starts burning (for example wood) or loses its mechanical strength (for example steel). Thereby, additional time is gained for fire defence (→ *fire resistant paints*).

fire protection: Complex of measures for the prevention of or defence against fire. In the coating industry, the fire risks are particularly high, above all because of the use of solvents. The main goal of fire protection is the prevention of combustible solvent vapour-air mixtures, using exhaust systems and avoiding sources of ignition like flames and → *electrostatic charge.*

fire resistant paints: In fire prevention (→ *fire prevention paints*), fire resistant and intumescent paints complement one another. While the purpose of intumescent paints is to reduce the negative results of a fire, which already exists, fire resistant paints render more difficult the ignition of a fire by means of their incombustible components. In textile fire protection, for example, high chlorinated incombustible components can be employed as film formers. These are generally combinations of polyvinyl chloride copolymers with chlorinated paraffins. However, the trend is move to halogen-free systems. These can be supported by the addition of water separating fillers (extenders), like aluminium hydroxide.

flaking: Flaking, as paint defect, can be caused by wrong substrate pretreatment like, for example, insufficient degreasing and/or an incorrect coating formulation (for example defective elasticity, or too little dissolving power to the substrate). Often, flaking shows the highest level of coating degradation, caused by weathering exposure.

flame cleaning: Method for the quick removal of rust and scale, above all from large steel structures. By means of a reducing flame (no oxygen excess), the metal surface is heated. Because of their different thermal expansion coefficients the rust and scale layers flake off and at the same time the rust is decomposed by reduction. After brushing off, the clean steel surface has to be immediately coated with a primer.

flammable: Characteristic of hazardous working substances, which is particularly important for the coating technology, because of the flammability of numerous solvents. In the national and international safety regulations, a classification of the flammable substances in three different categories is established according to their flash point. "Extremely flammable" are substances with a flash point below 0 °C, "highly flammable" those with a flash point below 21 °C to 0 °C and "flammable" those with a flash point from 21 °C up to 55 °C. Extremely flammable and highly flammable substances and their preparations have to be labelled with a flame symbol, while flammable have not. This is the reason why, in the development of coating formulations, one aims at reaching flash points which possibly are not below 21 °C.

flash off: Important intermediate step of coating application processes. Flash off means the time interval one has to wait before the recoating of an evenly applied coating layer, in order to permit sufficient volatilisation of the solvents. That can play an important role, when optimal levelling and prevention of sagging are required at the same time; in addition, relating to the optimal effect formation, in the case of metallic coatings, and to other properties of the coating surface. With regard to stoving systems or forced drying, flash off means the additional time which has to pass before the coated work piece is heated in the oven. If this time is not observed, one runs the risk of blistering and pinhole formation.

flash point: It is the lowest temperature, at which a test solvent generates a vapour-air mixture, capable to be ignited by an external small flame. The flash point is the base for classification of liquids in hazard classes, according to the Betriebssicherheitsverordnung BetrSichV (Operating Safety Ordinance) as well as to the → *GHS*.

flatting agents: Also called "matting agents". They are used to produce different gloss degrees in coatings. Matting simply means to roughen up the coating surface. The less a coating surface is smooth, the matter it looks. Matting is generated in coatings by means of solid particles, which due either to their particle size or to the filling rate of the coating material (→ *pigment volume concentration*) are not able to dip completely into the coating film. Therefore, they influence negatively the smoothness of the coating surface. In several cases, for example in dispersion paints, extenders are employed as cheap flatting agents. Yet, if very uniform and exact flatting degrees are needed, it is not possible to avoid the use of expensive flatting agents. These are either so called precipitated silicic acids, with very uniform particle sizes and high porosity, or synthetic waxes which also enhance the slipperiness and the blocking resistance of the coating layer. In some coating systems, flatting can be produced by the addition of only partially compatible film former components.

flexibility: Coating property, particularly important for coatings, which during the processing of the coated work pieces are subject to strong deformations. Typical examples are coil coatings, tube enamels, deep-draw coatings. To evaluate flexibility, there exist different standard methods (→ *rod bending test*).

floating: Phenomenon, which can only appear in coatings containing several pigments. There, in the still wet coating film, a separation of the pigment particles takes place due to their different settling rate. This separation is the more visible the more different the densities and the particle sizes of the involved pigments are. Because of the vertical streams developing in the wet film, which more or less intensely drag along the different pigment particles, floating can show negative results at the coating surface such as non-uniform colours, Bénard's cells or stripe formation. The addition of special additives can mitigate such effects (→ *anti-floating agents*).

flocculation: Effect of an unstable dispersion condition of pigment particles in a coating material. If the primary particles and the aggregates, which are generated during the → *dispersion process*, are not sufficiently wetted or if the wetting agent layer, absorbed on the pigment surface, is incomplete or the wetting agent has been dissolved by inappropriate addition of solvents (→ *solvents shock*), the forces of attraction between the pigment particles are reactivated. The result is the back-formation of conglomerates, which however, because of the inclusion of binder in their hollow spaces, can be reduced to small pieces easier than the original agglomerates and therefore, for linguistic differentiation, are called “flocculates”. The flocculation is one of the main causes of settling, gloss reduction and floating. Yet, in some cases a controlled flocculation (co-flocculation) among different pigments is provoked voluntarily by the addition of suitable wetting agents, in order to prevent the formation of visible colour differences, due to floating or the hard settling of fillers (extenders).

floor varnishes: Coating materials, which are characterised by high wear resistance and abrasiveness. Abrasiveness is necessary because of the danger of slipping and is achieved by the use of very hard fillers (extenders) like silica flour. The employed binders are also wear resistant and, as a rule, chemical and oil resistant as well. The most important systems are chlorinated and cyclised rubber, polyvinyl chloride copolymers, epoxy systems and two-pack polyurethane systems. Regarding the coating of parquet elements, UV- and electron beam curing systems are increasingly used.

flop: Term used for the change of brightness of metallic coatings at different angles of observation. The flop is an important tool for the evaluation of the brilliance and the colour of metallics. It is based on the fact that to an observer an object of metal looks the more metallic the bigger the contrast is between the bright gloss zones und the dark shadow zones, which lie at close quarters. For the determination of the flop, the coated panel is first held at an angle of 45° to the incident light at eye level. At this angle, the metallic colour has its highest brightness, because the light rays, which are reflected from the aluminium platelets, almost completely reach the eye. Then, the panel is inclined so that, due to the new angle of reflection, the rays pass by without reaching the eye. Therefore, the colour assumes its lowest brightness level. The bigger the brightness difference is between both angles of observation, the more brilliant the metallic colour looks and vice versa. The flop is influenced on the one hand by the particle size and the quality of the employed aluminium bronze, and on the other hand by the incorporating process of the bronze into the coating and by the coating application. The finer the used bronze is, the lower is the resulting flop. Fine bronzes give also less brilliant hues than coarser ones. During the incorporating of the bronze into the coating, one has to pay attention that the aluminium platelets are not damaged by too high shear forces. This would increase the fraction of fine particles and therefore it would cause a greying of the bronze. Finally, the application influences strongly the optimal orientation of the aluminium platelets in the coating film and therefore also the flop. The less perfect the horizontal orientation of the platelets is, the smaller is the light amount, which at the 45° angle reaches the eye, resulting in lower brightness. Conversely, by augmenting the inclination of the coated panel the amount of scattered light increases and the colour becomes less dark. In addition to the real brightness flop, some metallic colours, also show a pronounced "colour flop" due to the kind of pigments employed, that is the change of the hue depending on the angle of observation. Because of both flop effects, the colourimetric evaluation of metallic colours is only possible by means of special spectro photometers ("gonio-photometers"), which are able to determine reflection curves at different measurement angles.

Florida test: Variant of the outdoor weathering, which long ago became an international quality standard, above all for high performance coatings like automotive OEM and refinish coatings, aviation coatings etc. The particular climate conditions in Florida cause the most important damaging weather factors for coating materials. These are: high air humidity, heat and high UV-radiation. Through this combination, alterations in the coating are visible much sooner than, for example, in Europe. A rule of thumb is that a year Florida test corresponds to about two years weathering exposition in Europe (→ *weathering station*).

flow-coating: Application method, which consists of pouring the coating over the work pieces, mostly in a closed system. The coating material excess is collected and reprocessed. In a modern variant of this process, the coating is sprayed onto the work pieces by means of large spray nozzles and also recollected in a basin. The last method is particularly suitable for large and bulky objects.

flow cups: They are the most frequently employed devices to determine the → *viscosity* of coating materials. Flow cups are extremely easy to use and they can be applied every where. There are different types: DIN, ISO, Ford cup, but they all are based on the same principle. The cup of light metal has a standardised capacity and a nozzle in its lower end. The nozzle can be kept shut with a finger or with a metal stopper, while the coating is poured into the cup up to its upper brim. Afterwards, the nozzle is set free and simultaneously a stop watch is started. The time, which the coating material takes to flow completely out of the cup, is recorded in seconds. The flow cups are normally connected with a thermostat, since adherence to the prescribed temperature is important for the precision of the measuring.

flow properties: → *rheology*

fluidised bed process: Rational application process for thermoplastic powders (→ *powder coatings*), which is technologically situated at the boundary between powder and plastic coating. In a fluidising vessel the powder is whirled by means of an air stream, coming from the bottom, in order to form a uniform fluidised bed. The work pieces to be coated are heated in an oven to temperatures above the melting temperature of the employed powder. After that, they are dipped into the fluidised bed where the powder particles, which fall on their surface, fuse immediately and form a coating. The film thickness depends on the dipping time and on the object's temperature. With this application methods, filigree parts, like racks for refrigerators and dishwashers as well as wire baskets for baths and wash-houses, but also fittings can be provided with very thick and uniform layers, which are mostly based upon polyamide powders.

fluosilicate treatment: Method to prevent efflorescence phenomena, caused by calcium salts on mineral substrates. The calcium salts, which in the course of time are dissolved from deeper layers and transported to the surface, form on the surface of the masonry an ugly white patina. To prevent it, masonry is impregnated with hexafluosilicate solutions, which turn the calcium salts into non-soluble compounds.

foam: It is generated in general by all stirring processes, during the coating manufacturing, in which inevitably an indefinite amount of air is incorporated into the coating material, with the exception of working with a → *vacuum dissolver.* Foam is nothing bad, provided that it is not stable and bursts spontaneously. However, several coating materials, above all the water-borne ones which often contain emulsifiers similar to soaps or other surface active substances, tend to form stable foams which would stay in the system. In such cases, it is necessary to take remedial measures by the use of → *defoaming agents.*

forced drying: Acceleration of the reaction velocity of two-pack systems or also of the drying of coating systems, due to solvent/water evaporation by means of the temperature. In practice, this takes place at 60 °C to a maximum of 80 °C. Higher temperatures can cause problems like blisters, pinholes etc.

formaldehyde: Raw material for the production of important film formers which are used as combination partners for stoving systems. These are phenolic, urea and melamine resins. Formaldehyde is bonded to phenol, urea or melamine by an addition reaction. This generates reactive methylol groups, which later by stoving are able to cross-linking with other film formers containing OH-groups (polyesters, acrylates etc.).

formulation: The coating formulation contains almost the complete knowledge of a coating material, because it assembles the results of long lasting research, development and testing work. In addition, it represents the result of the constant optimization of the manufacturing process, with the goal to produce the coating material as cost-efficiently as possible and complying with high quality standards. Therefore, the traditional slogan that coating formulation is the "treasure chest" of a coating factory, does not sound exaggerated. Coating formulations are structured in process stages which are called differently, depending on the coating factory. The raw materials, the semi-finished products and their amounts are listed, according to the order of employment. Furthermore, they are subdivided into groups, which correspond to single process stages, and are provided with work instructions concerning their handling. Such work instruction could be, for example, "Premix 1 and 2 in a batch vessel; add slowly by stirring 3, 4 and 5". Moreover, the formulation contains exact information about the dispersing units and the parameters to be used. Formulations of pigmented coating materials are normally divided in two parts and consist of a → *mill base formulation* and the → *let down*. At some points in the formulation, there are indications of the tests, which have to be carried out, and the corresponding specified values, which have to be reached before one is allowed to go to the next process step; for example: "grindometer value under 20 μm" (→ *grindometer*). Finally, modern coating formulations are provided with extensive information about protective equipment and procedures for environmental protection and waste disposal.

formulation development: Development and optimisation of formulations for the coating manufacturing. Formulation development is the most demanding and extensive task of the coating design and realisation process, because in the coating formulation converge not only the results of many serial experiments and application tests but also a lot of know-how, process engineering as well as environmental aspects. Formulation development requires extensive knowledge of the raw materials and semi-finished products to be employed. In order to make the right choice of the formulation components for the single process stages, not only it is necessary to know how these can react with each other, or under the influence of factors like temperature and shear strain, but it is also important to know of their wetting, solubility and compatibility behaviour. The result of formulation development is the coating formulation (→ *formulation*), which contains a qualitative as well as quantitative list of the coating raw materials and semi-finished products to be employed. It represents a complete work instruction and process description to manage the manufacturing of the coating material.

free-flowing properties: Important technological property of powder coatings. The processability of the powder coatings (→ *EPS coating* and *fluidised bed process*) depends largely on the free-flowing properties, because the powder application requires an optimal homogenization with the transport air as well as the fluidised bed air (fluidising properties). In the past, free-flowing properties were evaluated by pouring out a defined cone of powder with subsequent measurement of the cone angle. Nowadays, one uses ever more a fluidising funnel with a nozzle and measures the flow time of a defined powder coating amount. This method is outstandingly suitable to evaluate the storage life of powder coatings, which does not depend only on the time but also on the storage temperature. Exceeding the storage life of powder coatings results in a significant increase of the flow time or even an obstruction of the nozzle of the fluidising funnel.

full tone colour: Sometime also called mass tone. Colour, which is generated by pigmentation of a coating material with only one pigment. Thus, it shows the original hue of the pigment dispersed in the corresponding application medium, without any effect caused by the addition of other pigments.

furnace carbon black: Relatively coarse carbon black, with a particle size ranging from 50 to 200 nm. It is produced by the burning of oils containing aromatic hydrocarbons, in an open pane with deficiency of oxygen. It is mostly used in tinting bases or mixing systems to match bluish grey colours.

furniture coatings: Materials for the coating of furniture, furniture parts or for their restoring. One has to differentiate between the manual and the industrial furniture coating. For the manual application, coating materials based upon nitrocellulose are still often used, because they dry very quickly, are outstandingly suitable to polishing and can be processed by craftsman techniques (for example as flat rubbing paints) to different flatting degrees. This is even more suitable for furniture restoring. Some time, twopack clear coats and acid curing systems are also employed. In the industrial application, which is mostly carried out by roller and curtain coating, UV-curing systems, especially those based upon unsaturated polyesters which harden in a few seconds, as well as two-component polyurethane systems dominate.

G

gas chromatography: It is the most frequently employed method to analyze solvents. The solvent is injected into the injector block of the gas chromatograph and thus evaporated. The vapours are incorporated by an inert gas stream (normally, helium) and transported with a constant velocity along the inside of a glass capillary column, which, due to its length of 40 up to 60 m, is spirally wound up. The inside of the column can be coated with different substances. Depending on the polarity of the substance, the solvent vapours, on their way to the end of the column, are differentially slowed down, so that they leave the column after different periods of time (retention times). Then, they are evaluated chronologically and quantitatively by means of a detector. The chromatogram, generated this way, is compared automatically by a computer with the memorised standard chromatograms of several solvents and evaluated accordingly. The gas chromatography allows for outstandingly reliable and quick information also when the smallest specimen quantities are used. Therefore, today it has almost completely replaced other evaluation methods like the distillation curve, density and refractive index.

gassing: Unwanted phenomenon, caused by a chemical reaction of coating components with each other or with traces of foreign matters. Most frequently the gassing appears in water-borne metallic base coats and zinc dust paints. In the first case it depends on an insufficient stabilisation of the aluminium platelets, which therefore can react in an alkaline medium with formation of hydrogen. Also in the case of zinc, a reaction with traces of water takes place and hydrogen is likewise generated.

gelling power: Property of the so called primary plasticisers, which are used to make more elastic brittle, physically drying film formers like cellulose nitrate, chlorinated and cyclised rubber, polyvinyl chloride etc. In doing so, the plasticisers act like "unsuitable solvents", which are not really able to dissolve the film former but which can only turn it, often by aid of heat or addition of real solvents, into a gel (gelling). Through physical interactions the plasticisers are fixed so firmly on the film formers molecules, that ideally they do not leave the system any more and produce their plasticising effect in a lasting way (non-migrating plasticisers).

German Standards Institute (DIN): Accredited central office, responsible for the issue of German Standards (DIN-Normen) in the fields of science, technology, economy and administration. Through its membership in the ISO (International Organisation for Standardisation), DIN aims also to support the international harmonisation of standards. Therefore several former DIN-standards have now changed into →*ISO-standards*. Standardisation work is carried out by different Standards Committees. For the coating sector the Standards Committees responsible are those for paint materials and similar coating materials as well as those for pigments and fillers (extenders).

GHS (Globally Harmonized System of Classification and Labelling of Chemicals): This system seeks to globally harmonize the classification of chemicals by type of hazard, and proposes harmonized hazard-communication elements, including labels and safety data sheets. First published by the United Nations in 2003, it has been implemented in the European Union as the CLP Regulation (Classification, Labelling and Packaging of Substances and Mixtures). Since 2015, the CLP Regulation has superseded all other regulations on the handling and transportation of hazardous substances and goods. The CLP classifies chemical substances and mixtures according to the type of danger that they pose, i.e.
- Physical
- Health
- Environmental

The chemicals are labelled with various hazard symbols (pictograms) and specific code numbers representing the corresponding hazard statements and precautionary statements. In addition to the pictograms, two signal words have been introduced: "Danger" to describe a high danger level, and "Warning" for less harmful chemicals.

glazes: → *wood glazes*

gloss: Optical property of surfaces, which is caused by the reflection of rays of light on their interface with the air. The gloss is higher the more directed is the reflection, that is the more the rays are mirrored by the surface with the same angle. That happens in the case of very smooth surfaces. The rougher is a surface the bigger is the portion of light rays scattered in all directions. This signify a loss of gloss. Therefore in the coating industry →*flatting agents* are used to generate all kinds of gloss degrees, from silk gloss to dead flat, by producing a specific micro roughness on the coating surface. Conversely, high gloss, as it is for example demanded by the automotive coating, can only be obtained by optimal levelling and by avoiding any negative influence on the surface smoothness, which could be caused by insufficient dispersion or pigment flocculation.

gloss measurement: It is carried out by means of reflectometers. These devices are able to detect the amount of rays of light, reflected from a coating surface, depending on their angle of reflection. According to the gloss degree ranges, different measuring geometries have to be applied. As a rule, the following measuring angles (calculated from the perpendicular) are employed: 20° for high gloss, 60° for moderate gloss and 85° for low gloss degrees. The devices have gloss standards to be calibrated before measurement.

gloss retention: Property of coating materials, which depends to a high degree on their light and weather resistance, and occasionally also on their chemical resistance. In most cases, gloss reduction is caused by a progressive degradation of the binder matrix, due to light and weather exposure, often supported by the photochemical activity of the pigments (→ *chalking*).

glycerine (propanetriol): It can also be called glycerin or glycerol. Trivalent alcohol, which still plays an important role in the polyester production, even though the trend goes more and more to trimethylol propane. Since glycerine is a main component of the natural fats and oils, it is used above all to manufacture → *alkyd resins* when they are produced starting directly from the oil rather than from the separated fatty acids (two-stage process).

glycol-ether-esters: Group of good solvents for the medium-polar field with high boiler properties. They are above all used in isocyanate hardening systems, in several stoving coatings and in metallic base coats. The most important types are methoxy propyl acetate and butoxy ethyl acetate (butyl glycol acetate).

glycol-ethers: Water soluble solvents, which play an indispensable role as coalescence agents in dispersion paints and as auxiliary solvents in water borne coatings. The most important types are methoxy propanol and butoxy ethanol (butyl glycol).

gold bronze powders: Platelet shaped effect pigments of zinc-copper alloys, which are mainly used for printing inks but occasionally also in industrial coatings and for applied arts. By changing the proportion of the alloy components, different effects can be produced, from pale gold to old copper. However, in the automotive coating they are not employed because they tend to "drown" (sedimentation) during their application, due to their high density. This can cause clouding. Also, they are not weather resistant, because the copper, during exposition, forms verdigris.

gold coatings: Stoving systems for the internal coating of packaging, which have to fulfil high requirements concerning chemical and food resistance. They consist of combinations of epoxy and phenolic resins and are exclusively processed as clear coats. Their name is due to the yellow colouration caused by the phenolic resin component, which during stoving at 180 to 200 °C changes to a golden colour.

grind gauge: Also called grindometer. Simple device to evaluate the degree of dispersion of pigments and fillers (extenders) in mill bases or in coating materials. It consists of a longish steel block, which along its longitudinal axis has a wedge-shaped indentation. Lateral to the steel block is a scale, which shows the correspondent depth of the indentation in µm. For the measurement of the grittiness a few drops of the coating material are put on the deepest point of the indentation. By means of a doctor blade the coating material is drawn in the direction where the depth of the indentation steadily decreases. After that the tester holds the steel block in an oblique position, at eye level, against the light so that the just applied coating stripe reflects the gloss light. This way, eventual coarser particles appear as silhouettes. The grind gauge value (grindometer value) corresponds to the scale value, where the first uniform accumulation of specks can be seen. The grind gauge value gives only information about the maximum agglomerate size; properties like tinting strength, transparency, purity of the colour cannot be directly evaluated. Thus the grind gauge value is only conditionally suitable as a criterion for product approval. However, it is an important instrument for control and adjustment of the dispersing parameters, during the coating production.

grindability: Particularly important property of putties and sanding surfacers. Since these products, due to their surface appearance, have generally to be sanded, their grindability contributes in an essential way to the total velocity of the coating process. Grindability is greatly influenced by the PVC (→*pigment volume concentration*), the →*packing density*, the filler combination and the cross-linking velocity of the coating material. In this regard, talc, as a particularly soft extender (hardness according to Mohs = 1), plays a dominant role.

grinding media: Spherical solids, which are used in ball, sand and bead mills to disperse the pigment agglomerates (→*dispersion process*). In the closed attrition mills, beads of zirconium or aluminium oxide are mostly employed. In rare cases beads of steatite or steel are also used. Their diameters vary between 0,1 and 3 mm. In the grinding compartment of the attrition mill, the beads are put in a circular movement by a rotor, together with the mill base. During this, different shear and impact forces become effective, which scatter the pigment agglomerates. When the mill base leaves the grinding compartment, the beads are held back by means of narrow separating slits.

grittiness: It expresses the state of dispersion of pigments and fillers (extenders), which can be evaluated by the naked eye in a wet coating film (→*grind gauge*). It corresponds to the size of not sufficiently dispersed agglomerates (maximum agglomerate size) and eventually also to outstanding coarse aggregates.

grounding: → *earthing*

H

hammer finish paints: Effect paints, used primarily for coating of tool boxes, machine cases, industrial measuring devices etc. The advantage of the hammer effect is to make less visible dirt and mechanical damages. The effect is generated by the addition of high molecular silicone oils, which, due to their incompatibility with other coating components as well as their low surface tension, float to the surface where they produce crater shaped cavities. Since the hammer finish paints contain also aluminium bronzes as effect pigments, the cavities appear like little dents, which are similar to those produced by a hammer delving into a metal surface.

hardener: It is the coating component, which is added to the "base material" of a two-pack system directly before the application to start the cross-linking reaction. Important examples of hardeners are isocyanates for polyurethane systems and polyamines for epoxy systems. Immediately after the hardener addition, the cross-linking reaction between the film former of the base lacquer and the hardener begins. During the reaction, the viscosity of the coating material increases until, after passing the so called "pot life", the material cannot be processed any more. In the coating language the peroxides, which are added for example to unsaturated polyesters in putties, are incorrectly called hardeners. Instead, these are only initiators of a polymerisation reaction and are not themselves involved in the chemical cross-linking.

hardness: Property of solids, which generally is described as the static resistance of their surface against an indenter. For many coatings hardness is very important, because coatings are often expected, among other things, to insure mechanical protection. Examples are the scratch resistance of automotive coatings and the abrasion resistance of floor varnishes. However, the optimisation of these properties is not only achieved by hardness but also by a certain degree of elasticity and slipperiness. To reach a compromise between hardness and elasticity in coating materials is often difficult, because hard films have the tendency to brittleness. Hardness of coatings depends in the first place on the cross-linking density as well as the physical forces of attraction between the film former molecules (→ *cohesion*), but it can also be influenced by the addition of fillers (extenders). Therefore, in putty formulations big amounts of talc are contained, because this soft extender lowers the hardness of the material and improves its grindability; conversely, in abrasion resistant road line paints rather hard extenders are used, like quartz and feldspars. Additives are generally not able to increase the hardness of layers. Nevertheless, special additives can make the coating surface less susceptible to scratches and other mechanical damages because they improve the slipperiness or the smoothness of the film. Thus, they contribute indirectly to the hardness of the surface.

hardness measurement: In the DIN and ISO standards, different methods for measuring the hardness of coating surfaces are described. The scratch or pencil hardness, the → *indentation hardness* according to Buchholz and the → *pendulum hardness* according to König are the most frequently used methods. In the case of the scratch or pencil method, hardness is evaluated by the resistance against the penetration of a pointed object, which glides on the surface, while in the Buchholz method the penetrating object lies on the coating surface. The pendulum hardness according to König serves to evaluate the elastic behaviour of a layer. This happens by the damping of the oscillations of a pendulum lying on the coating film.

harmful: Characteristic of hazardous working substances, which have to be labelled with the danger symbols "exclamation point" or "human torso" according to → *GHS*. This labelling is relative ly wide spread in the case of solvent borne coatings, since above all some solvents (for example aromates) but also certain film formers (for example isocyanates) and additives (for example biozides) are classified as harmful.

haze: In the coating technology this term is used to describe two different appearances. In the first case it means a milky opalescence in clear solutions, caused by either an insufficient compatibility of the coating components among each other, or a defective solubility in the employed solvents or the presence of micro foam. In the second case the result is a reduction of distinctiveness of image (DOI) in a dried film caused by scattered light of low intensity close to the main reflection. This variant of haze is also caused by the insufficient compatibility of the raw materials used, or by the wrong choice of solvent composition, for example if a non-solvent leaves the film as last solvent component. Other reasons can be a defective dispersion of the pigments or their flocculation, and disadvantageous temperature and humidity conditions during application. A reliable method for determining haze is the evaluation of the distinctiveness of image, also called DOI-value. This evaluation can take place either instrumentally or visually. The evaluation principle consists essentially of detecting a series of samples or letters with textures of varying degrees of fineness, which are reflected by a coated surface, and of determining their distinctiveness, recognition or readability.

heat stability: Important property of raw materials for the coating sector and of coating materials, which either during the hardening process or the common use of the coated objects are exposed to high temperatures. Thus, for example, no iron oxide yellow pigments should be employed in coil coating systems or powder coatings, because at the high stoving temperatures of 180 °C up to 220 °C these pigments would turn into iron oxide red with separation of bonded water. Instead, chromium and nickel titanium yellows are used for such purposes, which can stand temperatures of up to 1000 °C. For heat resistant stoving coatings, as a second example, expensive melamine resins are preferred, because the cheap urea resins show a clear tendency to yellowing from 130 °C onwards. Finally, for coatings with a steady exposure to high temperatures (internal coating of hearths, protection of exhaust pipes) high heat resistant silicone resins are used, which can stand working temperatures of up to 500 °C.

heavy metals: Metals, which show a high toxicological potential, if their compounds are assimilated by the human body. That is a main reason why in the last decades numerous coating raw materials, above all pigments and additives, were replaced by others. Examples of pigments, which by now have been fully or partly substituted, are the complete group of the cadmium yellow and red pigments (→ *cadmium pigments*), the → *lead chromates* and *molybdates*, lead white as well as the anticorrosive pigments → *minium* and → *zinc chromate*. Among the additives examples are the antifoulings on the base of mercury and lead driers. In this regard, it is important to specify that only the solubility of their compounds in the gastric acid is responsible for the dangerousness of the heavy metals. Chromium, for example, in the relatively easily soluble chromates is considered dangerous, while in the extremely hardly soluble chrome oxide green is reputed to be biologically harmless.

hiding power: Capability of a pigmented coating material to hide the colour or colour differences of the substrate. The hiding power of a layer depends to a high degree on the film thickness, the → *pigmentation* (pigment content) and the type of pigments employed. Since the contrast ratio of the substrate also plays a significant role in hiding power, today the trend is to apply coloured primer surfacers, especially when expensive and poor hiding organic pigments are used in the topcoats.

high boiling solvents: Solvents, whose boiling range is above 150 °C. They play an important role in the drying regulation because they keep the film "open" for a sufficient period of time, until all other volatile components can leave it without producing any problems. Besides they insure the improvement of the levelling. However, an overdose of such solvents can lead to defects like curtaining and retardation of the → *dust-free drying.*

high dispersed silicic acids: Also called "pyrogenic silicic acids". They are produced by flame hydrolysis from silicone tetra fluoride and represent one of the most important groups among the rheological additives (→ *rheology*). They consist of extremely fine silicone dioxide spherical particles (in the range of nanometers), whose surface is covered with innumerable OH-groups. The OH-groups are able to form with each other so called hydrogen bridge bonds, which are responsible for → *thixotropy* and → *pseudoplasticity.* Also hydrophobic types are available, which are made non-polar by coating with silane. The latter show a reduced thixotropy effect in comparison to the normal polar types; on the other hand they are less hygroscopic and so particularly suitable for isocyanate based systems or zinc dust paints, which are sensitive to water.

high pressure spraying: Widespread application method, which is used above all in the automotive refinish. It is also called "pneumatic spraying", because compressed air between 2 and 7 bar is used for the coating atomisation. As a rule, the spray guns are equipped with nozzles between 1.2 and 1.5mm. According to the position of the coating supply pot, one differentiates between gravity pot guns (supply pot mounted on the gun-top) and siphon pot or pressure pot guns (supply pot suspended below the gun). In the automotive coating OEM the guns are connected to a central pressure kettle. In the case of the gravity pot guns, the coating flow to the nozzle is produced by the weight of the coating and in the case of siphon pot guns by the partial vacuum generated at the nozzle. In the pressure pot guns, a portion of the compressed air is used to exert an additional pressure on the coating material in the supply pot. However, in all cases the coating atomisation takes place only out of the nozzle. Here the coating jet is seized by the air streams, which emerge from the air circuit round the nozzle, and immediately atomised. Two lateral so called "horn passages" provide additional compressed air, which serves to adjust the breadth of the jet. The main advantage of this application method is the outstanding atomisation effect, which allows most satisfactory application results. A partial disadvantage is the high → *overspray,* but which can be significantly reduced by electrostatic assistance (→ *electrostatic application*).

high solids: Usually called "high solid". Coating materials with a high solids content. Since this term is not standardised, the thresholds of the other classifications "medium solid" and "low solid" are somehow arbitrary and mainly indicated by practice. Concerning the classification of coating materials according to their solids content (non volatile content), the following rule of thumb is normally used: low solid = almost 20 up to 40 %; medium solid = almost 40 up to 60 %; high solid = almost 60 up to 80 %. The legal mandate to reduce the VOC-content in coatings (→ *VOC-directive*) in the last years has strengthened the trend to use high solid in application areas where waterborne and powder coatings cannot be used for technical or economical reasons. To reach the solids content in high solid systems, it is necessary to employ low molecular film formers and so called → *reactive thinners*. Particularly suitable to the high solid technology are isocyanate based two-pack systems, because their low molecular hardener components contribute much to the necessary solids content; but also two-pack epoxy systems, alkyd resin coatings and melamine based stoving systems can be used for this purpose.

hot dip galvanising: Very effective corrosion protection method for iron and steel. It is based on the fact that zinc, on conditions capable to promote corrosion, is dissolved instead of iron because it is less noble than iron. The zinc ions generated this way form by reaction with water and oxygen very tight layers with good adhesion, which protect iron against corrosion (→ *zinc dust primer*). In the hot dip galvanising the iron work piece is dipped into melted zinc. Thus, zinc layers are generated which can reach a thickness up to 80μm and, during the cooling down, form a characteristic crystalline bloom structure. This can be more or less fine, according to the different process variants. Hot dip galvanised surfaces are problematic as coating substrates because of the presence of cavities and cracks, above all on the interface steel/zinc.

hot spraying: Variant of the spray application, which in comparison with the conventional spray methods, makes it possible to reduce the solvent emission and at the same time to increase the application efficiency. In this regard, the coating material is held at a temperature between 60 and 80 °C during application, so that its viscosity reduces depending on the temperature. In this manner, it is possible to reach the necessary application viscosity with less solvent. A further advantage is the application of more solids content, and therefore more dry film, in a single working operation. Disadvantages are the insufficient levelling and the difficulty to have a soft overlapping of the spray jets.

house paints: Generic term for all coating materials, which are used to coat building elements of wood, metal or plastic. Among these products are in the first place coating materials, suitable to manual application with brushes or rollers (→ *decorative paints*), but also materials which are used for industrial applications, for example on window frames or doors, by means of dipping methods.

hue: One of three characteristics, necessary to describe a colour. In the colour measurement, hue defines the colour shade. The other two basic features of a colour are its lightness and its saturation (chroma).

HVLP: Variant of the high pressure spraying, which produces less → *overspray* and a lower loss of coating material due to modified pressure and coating supply conditions in the spray gun. HVLP means "high volume, low pressure", that is the air pressure is reduced and instead the air volume for the coating supply is increased. The gun supply pressure in the HVLP-application is about 2 bar, while the conventional high pressure guns work with 3 up to 5 bar. The atomisation by the HVLP-method is insignificantly worse, but on the other hand the application efficiency is clearly better in comparison to the conventional high pressure spraying. During the application of most coating systems, with the exception of some metallic materials, HVLP shows no disadvantages compared to the conventional high pressure spraying, if the application parameters have been adjusted very carefully.

hydrogen bridge bond: The strongest among the intermolecular forces of attraction. Due to its mechanism of formation it ranks between the pure physical forces and the chemical bonds. It is generated when a proton donator (hydrogen delivering group) and a proton acceptor (hydrogen receiving group) meet. As proton donators act mostly NH- and OH-groups and as proton acceptors nitrogen or oxygen atoms with free pairs of electrons. Thus a strong force of attraction develops between the hydrogen and the free pairs of electrons, which is clearly higher that the attraction of the → *Van der Waals-forces* or of the simple → *dipole forces*. Hydrogen bridge bonds are responsible, among other things, for the high polarity of the water and for its unusual boiling and solidification behaviour, also called "anomaly of the water", as well as for the polarity of the alcohols and the thixotropy effect of some rheological additives, for example of the → *high dispersed silicic acids*. Besides they play a decisive role concerning adhesion and cohesion in layers and at interfaces.

hydrogen ions concentration: → *pH-value*

hydrophilic: It means "water friendly". Property of many polar substances, above all those that because of their molecular structure, are able to generate hydrogen bridge bonds with the water. Such substances can be well wetted by water and form with it stable absorption layers.

hydrophobic: → *water repellent*

hydroxyl group: OH-group, which is, among other things, the functional group of the alcohols. It plays a very important role in the coating chemistry, because it is involved in many reactions, which either lead to the formation or to the subsequent cross-linking of binders. For instance, hydroxyl groups (OH-groups) react with acid carboxyl groups by the synthesis of polyesters, with isocyanate groups by the cross-linking of two-pack systems, with methylol groups during the hardening of stoving systems etc.

I

ignition temperature: It is the lowest temperature at which the vapours of a combustible liquid, in a mixture with air, ignite without an external flame (→*flash point*). The ignition temperature serves, among other things, to classify the solvents in ignition groups and is also used to define explosion-proof areas in the coating production, where the electrical devices and machines used in the process have to be properly equipped according to the relative explosion hazards.

illuminant: It means a defined variant of "white light". An optimal white light would consist of the sum of all spectral lights with absolutely equal intensity. This situation hardly occurs in practice. Sun light at noon, for example, differs a lot in its composition from evening light. In the evening light, the spectral red components are clearly more intensive. On the other hand, in a neon light the blue components dominate. Since the appearance of a colour depends also on the composition of the shining light, it would not correspond to practice to use an optimal white light for colour evaluation. Therefore, different illuminants have been defined in their compositions and standardised (standard illuminants), which are similar to those often found in practice (→ *metamerism*).

immission protection: Complex of measures to protect human beings and the environment against the effects of air pollution and other negative influences. There are many national and international regulations, which are increasingly harmonised. Particularly important for the coating industry is the → *VOC-directive,* which aims at a significant reduction of the solvent content in coating materials.

impact test: It is used to evaluate the rapid deformability of coating films and to complement the slow →*Erichsen indentation*. A steel weight, which is inside a leading tube, is let fall from different heights on the coated panel. Since the weight has a spherical end, it produces a circular dent on the panel. The fall height which has caused the first cracks in the film, is evaluated. One can let fall the weight on the coated side as well as on the reverse of the panel. The values of the impact test are mostly smaller then those of the Erichsen indentation, because in the case of a rapid dent formation the film has no time to equalise the strain.

impregnating varnishes: Varnishes, which are used to impregnate wire coils in the field of electrical engineering. They fill the cavities between the coated copper wires with an electrically insulating substance, as well as mechanically stabilise the coils. Stoving systems on the base of polyesters, polyvinyl formale, polyamides and phenolic resins are mostly employed (→*electrical insulation coatings*).

incineration, catalytic: Also called afterburning. Method to eliminate or reduce the immission of organic solvents in the atmosphere. By means of a catalyst the remaining solvents in the exhaust are burned as completely as possible to become water vapour and carbon dioxide. The catalyst has the function to accelerate the combustion reaction to such a degree that it may be concluded before the polluted exhaust is delivered in the atmosphere. The generated heat energy can be eventually recovered. Other widely employed methods are the thermal and the regenerative incineration.

incombustible: Property of substances, which have no or minimal tendency to participate in combustion reactions, that is in those with the air oxygen, whereby heat and light are produced. Most raw materials for coatings, with the major exception of above all inorganic pigments (carbon blacks are combustible!) and fillers (extenders) do not show this property. Since chlorinated hydrocarbons are almost no longer used, only water fulfils this condition as a solvent or thinner. This represents a main advantage for the employment of water borne coatings. In the binder and additive sector only highly chlorinated compounds rank among the incombustible or the fire-resistant substances (→*fire-resistant paints*).

incompatibility: Insufficient predisposition of two or more substances to form with each other a stable mixture. Incompatibility is normally caused by an incorrect combination of the polarities of the involved substances, which leads to insufficient solubility or repulsion (→*polarity*). The results of incompatibility may be separations, turbidity or surface damages like →*cratering* or →*blushing*.

indanthrene blue: Organic blue pigment with a very high light fastness. From the point of view of the colour shade, it completes the group of the phthalocyanine pigments in the reddish blue range. Due to its high price, indanthrene blue is mostly used for metallic colours in the automotive sector.

indentation hardness: Method of hardness evaluation, which falls back to the measuring of the indentation depth caused by a defined test load in the coating layer. There are different methods of carrying this out, described in different standards, for example the Knoop hardness number and the Fischer universal hardness, which cannot be dealt with in this compendium. Only the indentation hardness according to Buchholz (DIN EN ISO 2815), which is the best known method, is described here in greater detail. An indenter of metal, with a defined mass and equipped with a cutting edge, is placed on the coating surface for a defined time. After removal of the indenter, the length of the indentation left behind is measured by means of a measuring magnifier. The degree of the hardness is the quotient of 100 divided by the indentation length in mm.

industrial coatings: Collective term for all coating materials, which are processed by industrial methods, with the exception of OEM automotive coatings and wood coatings. Industrial coatings are materials particularly for → *coil coating*, coating of plastic, can coating, electrical insulation coating and most powder coatings.

inert pigment: In the coating material it behaves chemically in a completely passive way and reacts neither with other coating components nor with the substrate. Inert pigments are employed when a lasting resistance against chemicals, water and air pollution is particularly important. They have to fulfil this condition, above all as components of top coats. Conversely, the anticorrosive pigments for primers, filling primers and wash primers are chemically active and their protective effects are based upon specific chemical reactions (→ *reactive pigment*).

infrared drying: Often also called IR-drying. The heat transfer is caused by electromagnetic waves, having longer wavelength than the visible light, which are absorbed by the coating layer and the substrate. As a result, the radiation is turned into heat energy. Unlike the convection drier, where the heating is carried out by means of circulating warm air from outside inwards, in IR-drying the heating takes place preponderantly from inside outwards. This permits a more uniform solvent evaporation and quicker drying and hardening progress. Besides, IR-drying makes possible to warm up quickly, but also to cool down promptly, the coated objects. Thus substrates which are sensible to heat, like wood and plastic, can also be coated without too long heat exposition. The radiators used for this process are subdivided into short-wave, medium-wave and long-wave ones and altogether they correspond to a wavelength range from 1,000 to 6,000nm. The shorter the wavelength of the emitted radiation, the hotter and quicker is the warming up and the higher the energy transfer.

inhibitors: Additives, which are able to slow down the progress of certain chemical reactions or to retard their start. Practically their effect is the contrary of the →*catalysts*. In the coating technology inhibitors play an important role, concerning the stabilisation of coating raw materials and coating materials capable of polymerisation, like styrene and unsaturated polyesters, which otherwise would not be suitable for storage. Sometimes they are also employed in little amounts, to regulate the application properties of such materials (→*pot life*).

inhibitory effect of oxygen: Problem, which can arise by curing reactions according to the mechanism of radical polymerisation. Due to the combination of the produced oxygen radicals (→*radical*) with the polymer, the inhibition by oxygen impedes a complete hardening of the coating surface. This result makes itself felt by forming a sticky layer. In the case of unsaturated polyesters this problem can be solved by the addition of paraffin wax to the polyester, which, during the cross-liking process, floats to the surface and impedes there the contact between oxygen and the coating, or by the use of expensive so called “gloss polyesters”, which are hardly susceptible to inhibition. In the →*electron beam curing*, non-reactive protective gases are employed to displace the oxygen.

initial drying: First stage of the physical drying of a coating material after its application. A sufficiently quick initial drying can be important for the subsequent application steps (→*flash off*) and as a quality aspect concerning the sensibility of a coating material to contamination as well (→*dust-free drying*).

inorganic chemistry: Branch of chemistry, dealing with substances which are found in the world of minerals and stones (inanimate nature) or are derived from them.

inorganic pigments: Pigments of mineral origin or the types which are produced synthetically but that in their structures are similar to substances found in nature. Inorganic pigments are mostly oxides or mixed-phase oxides like iron oxides (yellow, red and black), titanium dioxide, cobalt blue, nickel and chromium titanium yellow, chromium oxide green; but also chromates and molybdates, like chromium yellow and molybdate red, belong to this group of colourants.

interfacial tension: Tension, which is generated at the interface between a liquid and a solid. It depends on the polarity of the phases involved and, together with the surface tension of the liquid and the solid substances, it is responsible for all wetting processes in coating materials (→ *wetting*). As a rule, the wetting of pigments succeeds the better, the lower is the interfacial tension between the binder solution and the pigment surface. The addition of wetting agents serves, among other things, to reduce the interfacial tension. Similar processes are also valid concerning the wetting of plastic surfaces by coatings.

interference pigments: They imitate the soft, slightly opalescent colour of mother-of-pearl, therefore they are also called "pearlescent pigments". In the coating technology they are mainly employed in the formulation of metallic colours for automotive coatings (so called "pearls"). The pearlescent effect is generated by light interference, caused by very thin layers of metal oxides with a high refractive index, as a rule titanium dioxide, which are coated on a transparent carrier platelet, consisting mostly of the natural silicate "mica". The basic principle of light interference is, that inside of this multi-layer-system, by means of multiple refraction and reflection, the light rays are reflected from different depths. As a result, depending on the thickness of the metal oxide layer and of the respective wavelength of the light ray, a displacement of the wave path may take place. Rays with synchronous wave paths become stronger by interference, those with non-synchronous wave paths extinguish. Thus, starting from white light, a chromatic interference colour is produced and reflected from the pigment surface. The rays, which partly go through the transparent pearlescent pigment, have the complementary colour to the surface colour and, depending on the reflection power of the substrate, the surface colour and its complementary colour mix with each other more or less intensely. As a result, colours arise, which show shades changing between two complementary colours depending on the angle of vision; for example Red-Green, Yellow-Blue, Blue(greenish)-Magenta etc.

interior can coatings: The oldest systems for the interior protection of cans are the so called →*gold coatings* on the basis of phenolic and epoxy resins. They are stoved at high temperatures (180 up to 200 °C) and form films with high chemical resistance, which are also suitable to sterilisation and therefore widely employed to coat cans for food. Additionally, polyurethane stoving coatings with blocked isocyanates are used for the interior coating of petrol containers and special polyvinyl chloride copolymers for the protection of cans for beverages.

interlayer adhesion: Adhesion between different coating layers. It can be improved by the enlargement of the contact surface between the coating layers by means of sanding (→ *intermediate sanding*). In case of wet-on-wet-systems (→ *wet-on-wet-application*) the interlayer adhesion is regulated by the addition of platelet-shaped extenders (for example china clay) or by the use of suitable solvents, able to dissolve slightly the upper layer of the substrate.

intermediate sanding: Pre-treatment of an already dried coating film by → *sanding*, before the application of the next coating layer. The need for intermediate sanding may have different reasons. In the case of wood coatings, intermediate sanding serves, among other things, to cut the wood fibres which have sprung up after the first coating application, in order to get an uniform coating surface.

ion exchanger: Device to eliminate salts from the water. The deionised water so produced is very important for the whole chemical technology as "process water". In the coating industry, considerable amounts of deionised water are used for the production of water-borne coatings. There are two variants of ion exchangers, cation and anion exchangers, which are steadily employed in pairs. The exchangers contain special plastic spheres, which are able to separate their own ions and in return to assimilate external ions, deriving from the dissociation of salts in water. In cation exchangers, hydrogen ions of the plastic are exchanged against positive ions (cations), contained in the water. In anion exchangers, an exchange of hydroxyl ions against negative ions (anions), contained in the water, takes place. Since hydrogen ions and hydroxyl ions are components of the water, the exchange automatically generates pure water. When the surface of the plastic spheres is already covered with external ions, these have to be regenerated by a special treatment.

ionic and non-ionic: Classification criteria for dispersing agents. "Ionic dispersing agents" are able to produce ions, that is electrically charged particles, by dissociation. One differentiates between "anionic", "cationic" and "amphoteric" types. In the case of anionic variants, quite big, negatively charged particles (anions) are generated, which derive mostly from organic metal soaps, sometimes also from phosphates and sulphonates. The cationic ones often contain amino groups, which are turned into positively charged, substituted ammonium ions (cations) by the assimilation of protons. Finally, the amphoteric ones are able to generate anions as well as cations, depending on the application medium. "Non-ionic dispersing agents" generate no ions. In their molecules are contained polar groups, at regular intervals, which act as anchorage points for wetting processes.

The most important examples are derivatives of ethylene oxide, which show numerous ether groups. In the broadest sense of the definition , the so called "wetting resins" also belong to this group, although they have rather the character of polymeric substances.

iron blue: → *Berlin blue*

iron oxide pigments: They are the oldest pigments in the history of civilisation. They have been mined for millennia as minerals and are still employed for artists' colours. Instead, in the coating technology only synthetically produced types are used. The following variants exist: iron oxide yellow FeO(OH), iron oxide red Fe_2O_3 and iron oxide black Fe_3O_4. Only iron oxide red is thermodynamically stable and heat resistant up to about 800 °C. Iron oxide yellow contains in its molecule water of crystallisation and therefore it transforms slowly into iron oxide red, already above 140 °C temperature. In the case of iron oxide black this happens above 180 °C. Otherwise, all iron oxide pigments have a lot of outstanding properties. They are extremely light fast and weather resistant, very cheap, have a good hiding power and, because of their high ultraviolet absorption power , they offer effective protection to the layers. A disadvantage is the low colour purity of iron oxide red and yellow, which restricts considerably their employment in brilliant colours. Iron oxide red and yellow can also be produced as transparent pigments, which are widely used for metallics and wood glazes.

IR-radiator: → *infrared drying*

irritant: Characteristic of hazardous substances and their preparations, which in case of improper handling can cause reversible irritations to skin and eyes and itch. In the coating technology, above all amine solutions for water-borne systems and acid solutions for electrodeposition coatings and acid-curing systems rank among such substances.

isocyanates: They contain in their molecules very reactive isocyanate-groups (NCO-groups), which can enter into addition reactions, above all with OH-groups but occasionally also with amino-groups and water. Thus, in the coating technology they are employed preponderantly as cross-linking partners for binders containing OH-groups. During the cross-linking reaction, polyurethanes are generated, which show outstanding properties. Therefore, the most important technological application of isocyanates is their use as hardener in two-component systems, for example in the automotive coating. The so called "base material" and the "hardener" are mixed together, a short time before application, in a prescribed ratio. At this point the reaction starts and progresses at room temperature (→ *pot life*). But it is also possible to employ isocyanates in stoving one-component systems. For this, they have to be "blocked" before in their reactivity by the addition of certain substances. Only such "blocked isocyanates" guarantee that, during the storage, no undesired reactions take place. Only at the high stoving temperature the blocking agent is separated and the cross-linking reaction started. Such stoving systems are often employed in the powder coating area, for wire enamels, can coating and coil coating. A further application opportunity offers the so called "biuret reaction" with water. Thus, outstanding water resistant substituted poly-urea is formed. Such "moisture curing isocyanates" are used to coat wood, pasteboard and building substrates by way of very high humidity exposition.

isopropanol: It belongs to the group of the alcohols and as cleaning agent and solvent plays an important role in water-borne coatings as well as in high polar conventional systems, which have to dry quickly (for example wash primers). A disadvantage is its flash point below 20 °C.

ISO-standards: International standards, which are elaborated by the "International Organisation for Standardisation", headquartered in Switzerland. To this organisation belong 90 national standardisation boards (→ *German Standards Institute*), which work together in different "technical committees", with the goal to facilitate the international exchange of goods and services and promote international scientific/ technological cooperation.

K

kaolin: Also called "china clay" because it is the main component of china ware. Chemically kaolin is an aluminium silicate hydrate, whose particles have a marked platelet shape. From the natural mineral, by means of special preparation processes, kaolin is produced in very white and fine particle size ranges. Additionally, by processing at high temperatures kaolin can be dehydrated. The so obtained calcinated types are somewhat harder and whiter than the normal ones. In the coating technology kaolin is widely used as filler (extender), particularly where an efficient "barrier effect" is required and the likewise platelet shaped → *talc* cannot be employed because of its too high binder requirement and/or its too coarse structure. This is certainly so in the case of the automotive coating OEM, where kaolin in primer surfacers and electrodeposition primers is a main filler. Because of its white colour, its reinforcement effect and its chemical resistance, kaolin is also often employed in dispersion paints and in industrial coatings.

Kesternich test: Classical short-test method for the simulation of industrial climates by means of a sulphur dioxide atmosphere (DIN 50018 and DIN ISO 3231). In the past, major importance was attached to this test. Now it is considered too hard a test and not so informative. However, for quality comparisons, it maintains a significant role of guidance in the development of new coating products, above all if used in modified versions.

ketone resins: Reaction products of ketones, for example dimethyl ketone or cyclo-hexanone, and formaldehyde. They feature high brightness and good compatibility and are not saponifiable. Ketone resins are mainly employed as brittle resins in physically drying systems, predominantly in nitrocellulose lacquers, in order to improve their filling power, as well as their gloss and polishing properties.

ketones: Important solvent group of medium polarity. They show a broad dissolving power similar to that of the → *esters* and number with them among the strongest solvents. Ketones are particularly suitable for quickly drying nitrocellulose coatings, polyvinyl chloride copolymers and epoxy stoving systems. The most important types are aceton, methyl ethyl ketone and, as a high boiling solvent, cyclo-hexanone. Negative aspects of the ketones are their relatively high price and intensive smell.

kneaders: Dispersing units for coating materials with very high viscosity like putties and sealing compounds. They consist of a trough in which thick kneading arms of steel rotate, so that their profiles move one in the gaps of the other, similar to the kneading fingers of two hands. In the coating technology their importance has greatly diminished, because most putties are now dispersed in so called → *vacuum dissolvers.*

L

lacquer: The word is derived from the old Indian term Laksha and referred originally to the "shellac", a natural resin produced by "lac insects". Later it was extended to coating materials developed by the Chinese culture, which were composed of the drying sap of the Chinese rhus tree and natural colourants. After drying, these "lacquers" left behind glossy layers with decorative appearance and efficient protection effects. Still later the word "lacquer" was introduced in the European world as a synonym for coating materials, which, after drying, leave behind glossy, resistant and visible films, which fulfil the criterion of "finishing enamels". Today this term is prevailingly used referring to completely natural coating materials.

lead chromates: Lead containing inorganic pigments with high colour purity. The lead chromates cover a wide colour range (from bright yellow to orange) and, until a few years ago, they ranked with the most used yellow pigments, among other reasons because of their low price and good hiding power. However, the high lead content and the related health hazards have considerably reduced their importance. Nevertheless, significant amounts of lead chromate are still produced for coating purposes.

lead-free: Ever more frequent requirement on the part of customers in connection with coating formulations. Lead, a heavy metal, has a high toxicity potential. This danger is visualized by the obligation to label lead containing coating raw materials with "skull and cross-bones", according to the pertinent safety regulations. Some coating components, for example lead driers, were almost completely replaced by lead-free alternatives several years ago. More difficult is the situation relating to lead-free pigmentations in the yellow and red range. In this case, dismissal of cheap and pure coloured lead pigments, like lead chromate and lead molybdate, implies a considerable increase in the cost of pigmentation.

lead molybdates: Lead containing inorganic red pigments with similar properties to the lead chromates. Unfortunately, the similarity also extends to application restrictions, due to the physiological risks involved, and the corresponding problems relating to lead-free pigmentations (→ *lead-free*).

leafing bronzes: Common term for aluminium bronzes, which float to the coating surface. Their opposite can be considered the non-leafing bronzes, whose platelet-shaped particles do not float but are completely embedded in the coating layer. The different behaviour of the two bronze groups is caused by the coating of the aluminium platelets with so called grinding agents (lubricants) during their production (→ *aluminium bronzes*). Leafing types are normally coated with stearic acid. They form an uniform metal layer with high reflection power, which cannot be coloured by mean of pigments. Therefore, leafing bronzes are hardly suitable for the formulations of metallic colours. The main fields of use for leafing bronzes are high reflective coatings to protect containers that store combustible substances against the sun, in addition to decoration paints for the applied arts as well as for book and offset printing.

let down: Also called cut back. Stage of the coating production where the mill base, at the end of the dispersion process, is completed by adding the missing components of the coating formulation.

levelling: Capability of a wet film to spread so that the structure of the coating surface, produced by the application process, can be levelled as well as possible (→*spreading*). Thus, the result of good levelling is a smooth surface. Levelling depends on good wetting of the substrate, as well as on the viscosity of the coating material and on the uniform surface tension of the component contained in the coating.

levelling agents: Surface active additives, whose task is to improve levelling through the reduction of the surface tension in the coating. Most often, silicone oils (→*polysiloxanes*) are employed but also fluorinated compounds and acrylates. The working of levelling agents is based on the fact that, due to their very low surface tension, they are refused by the coating materials in which they are embedded. As a result, they migrate to the surface or they accumulate at the interface between the coating and the substrate. On the surface, due to their tendency to spread (→*spreading*), they form a sort of "stretched skin", which levels the unevenness of the coating. In terms of the substrate, they improve wetting which also influences levelling in a positive way.

lifting: Damage to coatings which may occur due to the wrong choice of material during the application process, or may be caused by an error in the application of the coating material. If an old coating layer, which is not sufficiently hardened (for example a relatively recently applied alkyd resin coating) is overcoated with a material containing particularly suitable solvents for this system (for example with a two-component material), the old film may be dissolved so effectively that it loses its adhesion to the substrate. The increase in its volume, caused by swelling, leads to wrinkling, which can be visible on the surface. A similar situation may arise by overcoating too thick film layers, which harden very slowly because of high solvent retention.

light absorption: It is one of the basic mechanisms to generate →*colour*. In the coating technology, this concerns in the first place the pigments. The white light (sunlight, neon light etc.) is a mixture of electromagnetic beams with different wavelengths. According to their wavelength, the single beams can produce different colour impressions and only their combination results in the colour "White". If these beams are isolated, one gets a range of different "spectral colours" like in a rainbow. The pigments in the coatings, with the exception of the white pigments and most of the extenders (→*light scattering*), are able to absorb definite ranges of the white light and transform them into heat. The spectral ranges left behind form the colour. Practically, the pigments work like "coloured light filters". A red pigment, for instance, filters the green spectral ranges from the white light and leaves the red ones behind, so that only these are visible. A yellow pigment absorbs the blue and leaves the yellow ranges behind. Finally, a black pigment absorbs almost completely the white light. Thus the colour "Black" means "no light". Which spectral light ranges a pigment is able to absorb depends on its chemical compound and its crystal structure.

light fastness: Special term, which in the coating technology is predominantly used to describe the light resistance of pigments and dyestuffs.

light refraction: Change in the speed of light when a light ray passes from a medium (for example air) into another (for example water), if the two media have different optical densities, that is different refractive indexes (→ *refractive index*). The change in the speed of light is made visible by a bend of the ray at the interface of both media.

light scattering: Multiple refraction and reflection of light rays, caused by particles with a sufficiently high → *refractive index*. The refracted and reflected rays, differently from the reflection at an even surface, have no definite direction. Rather, they go in all directions and consequently form a kind of light cover around the particles. Beside → *light absorption*, light scattering is the most important mechanism to produce colour by means of pigments. White pigments practically show only light scattering, that is they reflect the complete light spectrum. In blends with absorbing coloured pigments or with black, they also scatter the non-absorbed spectral ranges and lighten the colour (→ *whitening power*). Coloured pigments too are capable of scattering light. The precondition is a high refractive index and a particle size of more than 0,2 μm. In these cases, the scattered light is naturally not white but it has the characteristic colour produced by the light absorption of the pigment.

light stabilisers: Additives, which have the purpose, in clear coats and glazes, to protect the film former and/or the substrate against the negative effects of UV-beams contained in sunlight. Light stabilisers are subdivided into two groups: UV-absorbers and radical interceptors. UV-absorbers are able to absorb directly UV-light and transform it into heat. Radical interceptors neutralise the effect of radicals, which have already been generated during the degradation process and which would further accelerate the degradation. The two light stabiliser variants complete each other and therefore are mostly employed together.

light stability: Resistance of a coating against the effects of light exposure, especially ultraviolet light (UV-light). Most often, this property is tested together with moisture and heat, in order to get a total evaluation of light and weather stability. To this end, different variants of outdoor weathering (→ *Florida test*) and also accelerated test methods (→ *weathering, artificial*) are employed. For coatings which are exclusively exposed to light, there are also pure light- or UV-tests. Coating degradation, after a sufficiently long UV-exposure, is made visible first by yellowing, then by the increasing loss of gloss and last by cracking. The light stability of a coating system can be improved considerably by means of → *light stabilisers*. In the case of pigments and dyestuffs the term "light fastness" is generally used, instead of light stability.

linoleic acid: Unsaturated fatty acid with two double bonds, which represents an important component of several semidrying and drying oils. Its specific advantage is the relatively low tendency to yellowing and, at the same time, a sufficient drying velocity. In the form of soyabean oil, of which it is the main component, linoleic acid is widely used for the manufacturing of white decorative paints.

linolenic acid: It is the most important fatty acid component of linseed oil. Due to its three double bonds it is responsible for outstanding drying properties as well as for a marked tendency to yellowing.

linseed oil: In the past it was the most important oxidative drying oil, due to its low price and its outstanding drying properties. For centuries it has been largely employed in the production of colours for artists and oil varnishes. In the modern coating technology, just like the other vegetable oils, linseed oil is used predominantly for the manufacturing of → *alkyd resins*. Because of its marked tendency to yellowing, it has lost some of its importance in the field of finishing coatings. This is certainly the case for white decorative paints, where linseed oil is almost completely replaced by soyabean oil and tall oil. Conversely, for all kinds of primers and surfacers, for printing inks and above all for corrosion protection primers, linseed oil alkyds are still very important.

lithopone: Before the introduction of titanium dioxide, lithopone was the most important white pigment. It consists of a mixture of zinc sulphide and barium sulphate, which are produced in a common precipitation process, in the right quantity proportions. There are several types of lithopone with different contents of zinc sulphide. Since zinc sulphide is the real pigment component (barium sulphate is only a filler), the pigment quality increases with its content. Due to its good application properties and its favourable rheology, lithopone is employed in heavily filled systems, like putties and sealing compounds as well as in wall paints and floor varnishes.

long oil: → *alkyd resins, long oil*

lotus effect: → *nanotechnology*

low boiling solvents: Solvents with a boiling temperature below 100 °C. Low boiling solvents are indispensable where a quick initial drying of coating films is required after their application. However, a too high amount of low boiling solvents can also lead to levelling problems.

low pressure spraying: In the modern application technology it is largely identified with the → *HVLP-process.*

low-solvent: Coating materials, which contain less solvent than the normal ones (so called medium solid). Most water borne systems (among the dispersion paints there are also solvent-free variants) and the → *high solid systems* belong to the low-solvent class.

M

maleic acid: It is the most important unsaturated dicarboxylic acid for the coating technology. The presence of one double bond and two carboxyl groups in its molecule makes it possible to insert double bonds, which are susceptible to polymerisation, into polyester chains and consequently to produce so called "unsaturated polyesters" (→ *polyester resins,* unsaturated).

maleic resins: Synthetic resins, which are produced from→ *colophony,* → *maleic acid* and polyvalent → *alcohols.* They are characterised by brightness, light stability and ample compatibility. They are mainly employed in the field of printing inks and in combination with physically and oxidative drying coatings to improve their drying properties, body and gloss.

manual rust removal: Removal of rust without mechanical or chemical devices. In the case of non-uniformly corroded surfaces, which show local holes or cracks, for example in the pitting , contact or crevice corrosion, a wire brush must be used. Conversely, uniformly spread surface rust can be removed most effectively by means of dry sanding paper. For high performances relative to the corrosion resistance of the coating systems after application, a manual rust removal does not normally suffice.

maritime climate: → *salt spray test*

mass tone: →*full tone colour*

medium boiling solvents: Solvents, whose boiling temperatures are between 100 and 150 °C. The use of medium boiling solvents in coating materials contributes to the balance of their drying properties. This is above all important for coordinating a sufficiently quick initial drying and a satisfying levelling of the coating film.

medium oil: →*alkyd resins, medium oil*

melamine resins: They serve to formulate high performance stoving systems (→*urea resins*). In this group of film formers, melamine resins represent the top quality, since they are not only outstandingly light resistant but they can also resist much higher stoving temperatures than the urea resins (above 200 °C) without yellowing. Therefore, they are suitable to employment in the coil coating sector, where temperatures up to 230 °C can often be met. Besides, the high cross-linking density of the melamine resins allows, specially in combination with acrylic resins, the formulation of coating materials with outstanding weather resistance and colour fastness as well as good resistance against chemicals and solvents.

melting point: It is the temperature, at which a substance of homogeneous composition melts. In the coating technology, such substances are to be found infrequently. Much more frequently one has to do with materials, which show a →*melting range*. The melting point evaluation is generally carried out in a glass capillary, either by means of the apparatus according to Thiele or of more modern devices, which display automatically the change of the light refraction of the substance as a result of the melting process.

melting range: It is the temperature range, where a substance melts, which consists of different molecules. In the coating technology, this concerns above all the film formers. The resins, which are used as film formers, are generally mixtures of molecules, which can differ from each others in their shape and size. The film former properties have to be considered as average properties. This refers also to their melting behaviour, which is not characterised by an exact →*melting point* but by a melting range. Particularly suitable for a quick evaluation of the melting range of polymers is the Kofler-heating-bar, consisting of a heatable metal bar with progressive temperatures. After calibration, by means of a substance with a known melting point, the pulverised resin is spread along the bar, so that it is observed at which point the resin begins to melt and where it has completely melted.

mesh size: Side length of the quadrangular meshes of a sieve. In the coating technology, it is normally given in micro meters (µm) and it is a measure to indicate from which particle size onward substances can be kept back from the sieve. This information is particularly important for the →*filling in* the coating production, where the purification processes of the coating material also take place.

metallic coatings: Common term for all coating materials with more or less clear metallic appearance. The metallic lustre is generated first by the addition of metallic effect pigments, in which case , with the exception of printing inks and special decorative paints, almost exclusively aluminium bronzes are employed. Often, modern metallic colours also contain amounts of other effect pigments (for example interference pigments) and are mostly coloured by means of transparent pigments like organic pigments, carbon black or transparent iron oxides.

metallic effect pigments: Pigments, whose optical effect is caused by light reflection at the smooth surface of metal platelets (mirror effect). In this they differ from other variants of effect pigments like interference pigments, liquid crystals etc. (→*effect pigments*). The most important types for use in the coating technology are the "gold bronzes" known as copper/zinc alloys and the →*aluminium bronzes.*

metal, noble: Metal, which shows little tendency to oxidation. The more noble a metal is, the less is its inclination to turn itself into an oxide or a salt by means of the air components, that is oxygen and water, or of acids. Very noble metals, like gold or platinum, maintain their metallic lustre also after millennia, that means, they normally do not transform chemically. The electrochemical "normal potential" represents the measure of the more or less noble character of a metal. The value of this normal potential can be read in the →*electrochemical series.* The more positive the value, the more noble the metal.

metal, non-noble: In comparison to a noble metal, relative to the tendency to oxidation, a non-noble metal behaves opposite (→ *metal, noble*). In the →*electrochemical series,* the more metals are non-noble, the more negative are their "normal potentials". This is explained by an increasing tendency to undergo transformations by chemical reactions. An example is the reaction of the non-noble metal iron with the natural substances oxygen and water, which leads to rust formation. The prevention of this reaction represents one of the most important tasks of the coating technology.

metamerism: Phenomenon, such that when comparing two colour specimen by use of different →*illuminants,* different colour impressions are provoked. In the case of the presence of metamerism, it could happen for example, that two colour panels by exposure to the illuminant D 65 (average day light) look quite similar, while they appear completely different when exposed to the illuminant A (neon light). This is normally caused by the use of different pigments, which generate different reflection curves (→ *colour measurement instrument*).

methoxy propanol: It is chemically the methyl ether of propylene glycol. Due to its water miscibility, methoxy propanol (MP) is an important solvent for water-borne coatings. However, it is also used in polar solvent-borne systems.

methyl cellulose: Rheological additive for dispersion paints with highly thickening effect. It is often used either in combination with hydroxyethyl cellulose or instead of it. Both are also widely employed as wall paper glue. They prevent the sedimentation during storage of the large amounts of pigments and extenders contained in the dispersion paints, and they reduce the tendency to splashing during the application by roller.

methylene chloride: Among the chlorinated hydrocarbons, it is the solvent with the lowest toxicity. Therefore, unlike the others members of this group, it is still used in paint removers (→ *chlorinated hydrocarbons*).

mica: Lamellar, rock forming silicate, common in nature, which is contained for example in granite. It is found mainly in the form of muscovite as potassium aluminium silicate. A special feature of this mineral is that it can be easily split into thin, transparent and flexible platelets. These can be grinded to fine powders, which still maintain their platelet structure. Such mica powders are occasionally employed as crack bridging fillers (extenders) in dispersion paints and concrete coatings. Mica plays a special, technical role as a substrate in the production of pearlescent pigments (→ *interference pigments*).

micaceous iron oxide: Special filler (extender) for heavy duty corrosion protection (→ *corrosion protection, heavy duty*). It is a platelet shaped iron oxide with very pronounced barrier effect. Due to its coarse structure, it can only be used in thick structure systems. Micaceous iron oxide should only be processed by means of dissolvers, because the use of bead mills would destroy its platelet shaped particles.

migration: Negative phenomenon, which appears in the use of plasticisers or other coating components, when these are not sufficiently fixed in the coating material (→ *gelling power*) or are overdosed. As a result, the plasticiser migrate, in the course of time, to the coating surface where they generate a oily or sticky layer. At the same time, the film becomes more and more brittle.

mill base: The part of the coating formulation, which serves to disperse the pigments and the fillers (extenders). In addition to the solid coating components to be dispersed, the mill base normally contains only a part of the film former and of the solvents and the complete necessary amount of the dispersing additives (→ *wetting agents*). The blend of binder and wetting agents should have a wetting power as good as possible in relation to the employed pigments and fillers (→ *wetting*) and it is not allowed to react chemically or to be damaged on the dispersing conditions. Thus, the mill base is a pigment concentrate (master batch), where the viscosity and wetting conditions are adjusted so that the shear forces necessary for the dispersing process depending on the used dispersing unit can develop optimally. At the same time the mill base has to be formulated so that only the minimal amount of the coating material undergoes the expensive dispersing process.

milling: Improper but usual term for dispersion process. Technically, "milling" means a process of triturating or crushing compact pieces of a substance. Since the pigment agglomerates are not compact (→ *agglomerates*), the term "dispersion process" should be used.

milling facilities: → *dispersing units*

mills: Improper but usual term in the coating language for → *dispersing units*. In engineering, mills are used to crush or triturate compact pieces of material. Conversely, in the coating technology the goal of dispersion processes is to separate from each other existing little particles, which only stick together by means of physical forces of attraction. It is important to mention, that in other technological fields the dispersing units of the coating industry are often employed as real "mills". For example, in aluminium bronze production ball mills are used to reduce to small pieces aluminium gravel while nip-ball mills (co-ball mills) are employed in chocolate manufacturing. These processes have surely contributed to the confusion in the use of the technical terms.

Milori blue: → *Berlin blue*

minimum film-forming temperature: It is the lowest temperature, which has to be maintained during the application of dispersion paints. If the working temperature falls below this value, the fluidity of the polymer particles is not sufficient to produce a homogeneous layer after water evaporation (→ *coalescence agents*). The result is the formation of cracks and flaking in the film, which make the coat useless.

minium: Mixed oxide of the bivalent and tetravalent lead. The minium ranks among the oldest anticorrosive pigments and, until recently, it was often employed even on rusted substrates without previous rust removal pretreatment, because of its outstanding corrosion protection effect. While its high toxicity potential reduces considerably its possibilities of application, minium maintains a certain importance in heavy duty corrosion protection processes.

mixing coating: Component of a completely standardised →*mixing system*. It contains only a pigment and in terms of its colour properties (colour location and tinting strength) and its workability, this component has to be as reproducible as possible. This is necessary to permit the user to employ it in ready colour formulations, which are delivered from the coating manufacturer.

mixing system: Very user friendly method to produce each colour, even in little amounts and at any time, according to the customer request. For years, mixing systems have prevailed in the fields of automotive refinish and decorative paints. However, the automotive refinish systems are the most demanding when it comes to colour accuracy. The customer, that is the automotive refinishing shop or the decorative paint shop, gets from the coating supplier a mixing station with a definite range of mixing coatings, a computer controlled scale and a data bank of colour formulations, for example of all automotive colours being on the market or of RAL colours. On the request of the customer, the formulation of the colour is chosen from the data bank and, by means of the computer, converted into the necessary amount. According to the calculated amounts, the mixing coatings are weighed together on the electronic scale and then intensively mixed by means of a special shaking device (→*mixing coating*).

mixture with white: Method for optimal evaluation of colour shades and tinting strength of coloured and black pigments and for the colourimetric and visual determination of their influence in pastel tones. The registration of colourimetric data for →*colour formulation calculation* requires, besides the measurement of the full tone, also the measurement of defined mixtures of the coloured and black pigments with white, called standard depths of colour. Finally, it is common, for the evaluation of the light fastness of pigments, to expose to weathering and UV-radiation the full tone, in addition to mixtures with white. In the case of mixtures with white, pigment degradation is made visible more quickly by fading of the colour than in the full tone, due to the scattering power of the white pigment.

moisture curing: Cross-linking variant of isocyanates based upon a reaction with water (biuret reaction). Thereby, substituted poly urea molecules are formed and carbon dioxide is separated. Moisture curing isocyanates have been already partly cross-linked by a preliminary reaction. Since they have an excess of isocyanate groups, they are able, after application, to undergo a further reaction, by means of air moisture, until the final cross-linking degree is reached. Moisture curing isocyanates are mainly used to improve the resistance of mineral substrates (concrete, sandstone etc.) and of wood.

moisture-induced corrosion: Variant of the metal corrosion, provoked by the impact effect of moisture and air oxygen. Outstanding importance has the moisture-induced corrosion of the most employed metallic materials, that is iron and steel. The mixture of iron oxide and iron oxide hydrate, generated during the corrosion process, is called "rust". Unlike the oxidation products of several other non-noble metals, for example aluminium, which form a compact, well adhering and therefore protecting layer (passivating coating), rust is porous und poorly adherent, so that the corrosion process can take place reaching the complete destruction of the material. Corrosion protection of iron and steel represents the economically most important task of the whole coating technology. Zinc corrosion, which produces the so called "white rust", plays an important role in the coating processes. In corrosion protection zinc is employed in different ways, in order to protect iron and steel (→ *hot dip galvanising* and *zinc dust primers*). In unfavourable conditions, the zinc layer can be progressively oxidised and transformed, by reaction with water and carbon dioxide, into a mixture of zinc oxide hydrates and carbonates (white rust). Before coating application, such white rust layers have to be carefully removed, because otherwise the coating adhesion could be negatively influenced. Therefore it is necessary that the zinc film not be sanded through, to maintain the steel corrosion protection.

molecule: Littlest particle of a substance, which is composed of several atoms linked with each others by means of special chemical bonds (atom bonds).

molybdate red: → *lead molybdates*

monomers: Molecules, which serve as basic components for polymerisation reactions (→ *polymerisation* and *copolymerisation*). The choice of the monomers influences decisively the final properties of the polymerisation products.

MWC-value (maximum working conditions value): Also called "exposure limit", depending on the geographic area of its employment. It indicates which maximum concentration of hazardous substances in the form of vapours, dusts or noise in the work place should not be exceeded, in order to avoid dangers to health. The MWC-value is an average value, which refers to a definite number of work hours. In the coating industry, it represents an important indicator for the correct handling of solvents.

N

nail lacquers: Quickly drying lacquers on the basis of nitrocellulose, which often contain, beside coloured pigments, also dye stuffs and pearlescent pigments. Because of the required standard of physiological harmlessness, they are equipped with citric acid esters as plasticisers. As solvent, the quickly evaporating acetone is mostly used.

nanotechnology: Relatively new field of research and development with versatile possibilities of application. The goal of this technology is the formation of very regular surface structures in the nanoscale range (1 nanometre = 1 millionth of a millimetre). In this case, due to the extreme fineness of the particles and of the produced surface structures, in addition to the chemical composition of the involved substances, physical surface phenomena also play a decisive role. So it is possible, for example, to achieve surfaces with outstanding hardness or to produce layers with self-cleaning properties (lotus effect). Such effects are obtained on the one hand by carrying out syntheses, which imitate self-organising structures of the organic nature, and on the other hand by production of special polymers, synthesised from organic and inorganic components (sol-gel-process); the manufacturing of so called “composites”, made of different substances, is also possible.

natural resins: Before the development of the oldest synthetic resins, that is phenolic resins, several natural resins of vegetal or animal origin like dammar, copal, colophony and shellac were essential coating components. First of all, in oil varnishes they served to improve significantly the physical drying (→ *drying, physical*) but also the body, the gloss and the polishing properties. By means of special cooking processes, they could be partially chemically linked with the oil of the varnishes. This led to a better weather and water resistance. Later, they were largely replaced by synthetic resins for coatings, due to their superior technological properties. The only natural resin, which still plays an appreciable role in the coating field, is the → *colophony*.

needle-shaped: Form of appearance of some pigments and fillers (extenders). In the case of pigments, it is particularly characteristic of iron oxide yellows. Since the colour shade of an iron oxide yellow pigment depends on the proportion of thickness to length of the needle-shaped primary particles, this particle shape causes a certain susceptibility to dispersing processes. Breaking into pieces of the particles, due for example to a too sharp dispersion, may lead to dirty colours. In the case of fillers (extenders), needleshaped particles are rarely found.

Newtonian liquids: Liquids, whose viscosity (flow properties) only depends on the temperature. In this they are different from the non-Newtonian ones, whose flow properties, besides the temperature, also depend on the shear rate, that is the stirring conditions of the liquid (→ *pseudoplasticity* and *thixotropy*). Newtonian liquids are for example oils, solvents and water.

nickel titanium yellow: Inorganic yellow pigment, consisting of a titanium dioxide lattice, in which nickel atoms are incorporated. Due to the presence of nickel, the pigment obtains a relatively pure, yet somewhat pale, greenish yellow hue. It owes its increasing importance to formulations of leadfree pigmentations in the green-yellow range, where it is mostly combined with organic pigments (→ *chromium titanium yellow*). Because of its outstanding temperature stability up to 1000 °C, it is also employed in coil coating and powder coatings.

nitrocellulose: → *cellulose nitrate*

nitrocellulose combination lacquers: At the beginning of the industrial coating sector in the 1920s , they numbered among the most important coating systems. In the automotive industry, due to their quick drying, they made it possible to move from the slow craftsman production, which still employed the old oil varnishes, to the first serial production. They are composed of the very fast physically drying nitrocellulose (→ *cellulose nitrate*) and of non-drying, mostly short oil alkyds (→ *alkyd resins, short oil*). The alkyd resins fulfil the purpose of plasticising the nitrocellulose while at the same time they improve the wetting power, the body and the gloss of the coating material. Nitrocellulose combination lacquers still play an important role in wood coating and in some segments of industrial coating.

non-aqueous dispersions: Dispersions, which contain an organic liquid phase instead of water. In the first place, they are the so called NADs (non-aqueous dispersions), consisting of special acrylic resins which are emulsified in aliphatic solvents. Dispersions of polyvinyl chloride in pure plasticisers or in plasticisers with additions of mostly aliphatic solvents also belong to this group (→ *plastisols* and *organosols*)

non-drying oils: They are such oils, whose main fatty acid components contain in their molecules less than two double bonds (→ *oils*).

non-ferrous metals: This term is largely used for metals which are important as technological materials, first of all aluminium, magnesium, zinc, copper and their alloys. Strictly speaking, also galvanised steel, as a substrate for coating application, belongs to this group. Concerning coating adhesion, non-ferrous metals are problematic substrates and they have to be mostly pre-treated with → *wash primers*.

non-leafing bronzes: Non-floating aluminium bronzes (→ *leafing bronzes*), which are fixed within the coating layer and therefore can be coloured by the addition of transparent pigments. The absence of the tendency to float to the coating surface has to be attributed to the use of oleic acid as grinding agent, during the bronze production (→ *aluminium bronze*). This lubricant produces a non-polar aluminium surface, which can be easily wetted by most film formers and solvents. Thus, unlike the leafing types, the aluminium platelets of non-leafing bronzes are not pushed toward the coating surface. Non-leafing bronzes are widely used in base coats of two-coat systems (→ *two-coat metallics*) but also in industrial coatings and furniture lacquers.

non-solvent: Solvent, which is not able to dissolve a determined binder, even with the support of other solvents. Non-solvents (→ *solvent, real* and *solvent, latent*) can only be added to coating formulations in little amounts as thinners. In these cases, the compatibility of the coating material has to be considered, including after longer periods of storage. Besides, the non-solvent is not allowed to evaporate more slowly than the real solvents in the coating material, otherwise, during the film forming, precipitation of coating components and haze may take place.

non-volatile matter: Often also called "solids content". It consists of all coating components, which stay in the film after the coating drying. The non-volatile matter is evaluated according to ISO 4618. A determined amount of coating material is weighed in a metal bowl and dried in an oven at defined conditions (temperature and time). After having weighed the bowl again, the non-volatile matter in % can be calculated. In the case of stoving systems, the temperature for the evaluation has to be at least as high as the prescribed stoving temperature, otherwise the products, which are separated during the cross-linking reaction, would not be detected.

normal colour vision: The capability of an observer to see a colour "correctly" (→ *defective colour vision*). This presumes a perfect functional efficiency of the colour detectors on the retina of the observer (→ *colour*). The basis for the evaluation of functional efficiency are the sensitivity curves to blue, red or green for so called normal observers, which were experimentally determined in 1931 and are described as standard colour matching functions in the DIN 6164, section 2. Self-evident, normal colour vision is a precondition for all activities whose purpose is a reliable colour matching.

novolaks: Non-curing phenolic resins with linear structures, which are prone only to physical drying (→ *drying, physical*). In the coating technology they play a rather subordinate role and are some time used in barrier primers, insulating coatings and flexographic printing inks.

O

occupational safety: It includes all measures, which in respect of the handling of hazardous working substances are relevant to prevent dangers for human beings and facilities. In the coating technology, fire and explosion danger as well as the physiological effects of certain substances on the human body are the main dangers. Measures concerning occupational safety are written into numerous regulations and laws.

oil absorption value: It is expressed as the mass of linseed oil in gram per hundred gram pigment or filler (extender) and it is a measure for the → *oil requirement* of the pigment. In the case of pigments and fillers with relatively coarse particles, the determination of the oil absorption value gives important information for the formulation of the approximate composition of a mill base. The evaluation takes place by means of a linseed oil of standardised quality, which is let drop from a burette upon a defined amount of pigment, put on a plate of glass or marble. There, the pigment is kneaded by hand with the oil by means of a spatula. The evaluation is completed when a soft paste is generated, which can just be removed from the plate with the spatula without tearing or crumbling. Since the oil absorption value depends significantly on the kneading technique of the operator, the application of its result in practice cannot be considered exactly reproducible.

oil boiling: Chemical linking of a resin with a drying oil. The precondition for it is the presence (in the oil or in the resin) of a conjugated double bond system (two double bonds, which are separated from each other by a simple bond), with which an isolated double bond of the other partner can react. The oil boiling, which nowadays is not common, was carried out manually in so called "brewing kettles". The open brewing kettle was heated on a gas flame and the reaction mixture was stirred by hand until the resin melted and the liquid reached a certain viscosity.

oil content: Content of oil or fatty acids in an alkyd resin (→ *alkyd resins, short oil, alkyd resins, medium oil* and *alkyd resins, long oil*). The oil content is always given in the form of the ester of the glycerine (triglyceride content) also when, for the production of the alkyd resin, the pure fatty acid has been employed.

oil paints: Pasty dispersions of pigments in oils, containing driers, which are mainly used in art applications. In order to avoid premature drying, they are stored into airtight tubes.

oil requirement: In practice it is determined by the evaluation of the → *oil absorption value* and it expresses how much binder has to be absorbed from a pigment or filler (extender), in order to wet completely the particles and to fill all interstices between the particles. The oil requirement depends directly on the → *specific surface area* of the pigment but also on its state of agglomeration or of dispersion.

oils: Natural esters from glycerine and → *fatty acids*. They can be of vegetable or animal origin, but the vegetable ones are clearly predominant in the coating application. The fatty acid mixture, which is esterified with the glycerine, may vary according to the provenance and the conditions of growth of the oil. However, the mixture shows a characteristic main component, which defines the essential oil properties (→ *fatty acids, saturated* and *fatty acids, unsaturated*). According to the number of double bonds of their main component, the oils are subdivided into drying, semidrying and non drying types. The most important drying oils for the coating application are wood oil, linseed oil and of more marginal use also some fish oils. Among the semidrying oils, soya bean oil tall oil and dehydrated castor oil ought to be emphasised. The main non drying types are coconut oil, peanut oil and castor oil. Oils are employed in the manufacturing of alkyd resins.

oil varnishes: Coatings, which contain drying oils as the main film former component. In order to improve their initial drying, in the past they were mostly combined with → *natural resins*, then they were boiled with → *colophony* or phenolic resins to form the so called "artificial copal" (→ *oil boiling*). The through drying was improved by the addition of blown oils or → *stand oils*. Today oil varnishes have lost their importance because of the superiority of other coating systems.

oleophilic: → *water repellent*

oleophobic: → *hydrophilic*

oligomers: Classification for low-molecular polymers with a molecular weight below 2000 g/mol. Unlike polymers, oligomers are formed by the linking of relatively few molecules of → *monomers* by means of a polymerisation reaction.

one-coat systems: Coating systems, which reach the required technological properties by the application of only one coating layer (→ *topcoat*).

open pore: Special term, used in the wood coating application. It refers normally to wood layers, which have been generated by thin → *glazes*, when the pores, situated between the wood fibres, are not completely sealed by the coating material. Thus, a moisture interchange between the inside of the wood and its exterior can be maintained permanently. Strictly speaking, the so called "ventilation primer surfacers" also rank among the open pore systems, although the very high water vapour permeability of these wood primers is produced by their own porosity.

optical properties: Common term encompassing all aspects, which influence the appearance of a coat. Among these rank the gloss, the surface structure and the colour, including the effects which are caused by → *effect pigments*. In the automotive industry the optical perfection of the surface, including the accuracy of the colour and of the effect formation, is generally only called "appearance".

orange peel effect: Coating damage, which can be produced by material defects or application mistakes. The most frequent causes are the absence of high boiling solvents or levelling agents, too high → *thixotropy*, too high application viscosity, wrong spray gun adjustment and too high application temperature. In a few cases, for example in special structured coatings, the orange peel effect is produced on purpose.

organic chemistry: Branch of chemistry, which is engaged with the research, the development and the synthesis of compounds, whose structures are similar or identical to substances found in "fauna and flora". The basis of organic chemistry are the different hybrid states of the carbon atoms, which make possible extremely versatile possibilities of binding to other atoms. The favourite partners of carbon in organic compounds are in the first place hydrogen (aliphatic and aromatic hydrocarbons), followed by oxygen, nitrogen and sulphur. The number of the known organic compounds is immense (between 4 and 6 millions). The source to manufacture raw materials for organic synthesis is almost exclusively the petroleum industry.

organic pigments: Pigments with pronounced colour properties, which are generally structured in a complex way (→ *inorganic pigments*). Since their production often requires several synthesis stages, they may be very expensive. In their application properties they differ from inorganic pigments because of their brilliant hues, high tinting strength and low hiding power (high transparency). Due to the fineness of their particles, they are difficult to disperse and their pronounced surface activity causes a negative rheological behaviour of the pigmented pastes (high pseudoplasticity). As a rule, this fact renders possible only low pigmentation rates. These properties are rather problematic for plain colours, above all in the case of the replacement of cheap and high hiding lead containing pigments in the yellow and red range (→ *lead-free*). For use in metallic colours, organic pigments are indispensable because of their transparency (→ *metallic coatings*). Concerning light fastness and weather and solvent resistance, the diverse organic pigment groups act in very different ways. While → *azo pigments* are poor to mediocre in terms of their light fastness and solvent resistance (no use in metallic coatings), phthalocyanine pigments (→ *copper phthalocyanines*), but also the particularly expensive quinacridones, perylenes and polycyclic pigments, are in this regard outstandingly good.

organosols: Dispersions of polyvinylchloride or of its copolymers in plasticisers and organic solvents. The employed solvents are only permitted to develop their dissolving power onto the polymer at temperatures higher then 130 °C, because at these temperatures the gelling process, provoked by the plasticiser, begins (→ *gelling power*). During the gelling process, the solvents, which are in reality latent solvents, play temporarily the role of gelling agents, making easier the physical adsorption of the plasticiser by a partial solvation of the polymer (→ *plastisols*). After that, the solvents leave the so produced homogeneous film. Thus, unlike plastisols, organosols do not rank among the emission free systems. Since solvents accelerate considerably the gelling process, organosols are particularly suitable for coil coating application. The layers generated this way show outstanding resistance against water and all kinds of chemicals.

outdoor stability: Resistance of coating materials against weather influences. Among these rank in the first place light, humidity and temperature, but also several air pollutants like industrial exhaust gases or salt fog.

outdoor weathering: Common term for the exposure of coatings to climatic factors, which can affect coating materials or coating systems in their outdoor use. In the coating technology, outdoor weathering normally means the long term exposure of coated test panels or objects to defined climatic conditions like maritime climate, industrial climate and Florida weathering.

outturn sample: Stored sample of a batch of coating after manufacturing, which, in the case of customer requests or complaints, can serve for quality comparisons.

overcoatability: This term refers to two different aspects of coating application. First it means the general capability of an already applied coating layer to be overcoated with the same or another coating material without showing coating defects like for example → *lifting* or → *bleeding*. This property depends on the coating composition (general susceptibility of the employed binders and/or pigments to some solvents), but it can also be influenced by the drying time, the drying conditions and the age of the coating. The second case of overcoatability refers to the time-dependent capability of a film, which has been applied by brushing, to be recoated with the same coating material and the same technique without that the already dried film be damaged by the brush bristles or dissolved by the solvents. The data sheets and the labels of coating materials for the brushing application (for example decorative paints and floor varnishes), which as a rule have to be brushed at least two times, contain exact indications about the time which has to pass before brushing the second time.

overcuring: Cross-linking condition of a two-pack system, which is caused by the addition of an overdose of hardener. Overcuring can be the result of an application mistake (wrong calculation of the stoichiometric proportion, defective dosing device), but it can also be, on a limited scale, deliberate if in some cases of application particular properties of the film should be emphasized. As a rule, overcuring improves the hardness and the weather and chemical resistance of a film, but at the same time it decreases its elasticity and adhesion.

overspray: It is the part of a coating materials, which gets lost during the application process (→ *airless spraying* and *high pressure spraying*). The overspray amount does not only increase the process costs because of the loss of material, but it also represents a considerable burden for humans and the environment. Overspray can be considerably reduced by means of expensive exhaust technologies, such as the employment of electrostatic assistance, but also of improved application devices (→ *HVLP*).

overstoving: Curing of stoving coatings at temperatures, which are higher than the ones prescribed in the processing specifications. As a rule, one tries to avoid overstoving because, depending on the coating composition, it can cause undesirable changes in the coating layer (blistering, yellowing or colour change). However, in some processes (for example coil coating or powder coating) temperature fluctuations have to be expected. Therefore, overstoving resistance is a relevant property for coating materials, which have to be applied by such processes.

oxidation: Chemical process, which is important above all for metallic materials and is normally based upon a reaction of the metal with the air oxygen or acids. During the oxidation, the material properties change for the most part negatively, some time to the point of the complete destruction of the material (→ *corrosion*). But there are also exceptions, for example the formation of very well adhering and compact layers of aluminium oxide, caused by the effect of air oxygen on aluminium, or the creation of passivating layers generated by phosphoric acid on iron (→ *passivating*).

oxidative drying: Special variant of the chemical cross-linking, concerning the drying → *oils* or their fatty acids and the alkyd resins, which contain them. In the oxidative drying, air oxygen is installed as a “natural” crosslinking partner in the coating layer. This process takes place by the temporary formation of hydro-peroxides and their subsequent decomposition, whereby reactive radicals are formed, which lead to the cross-linking. The reaction is considerably accelerated by the employment of driers.

P

packing density: Term, which is mostly used for highly filled coating materials like primer surfacers, corrosion protection primers, putties and dispersion paints. It expresses how compactly the pigment and filler particles lie together in the dried film or how large the hollow spaces, filled with film former, are between them. Theoretically packing density can be calculated from the → *pigment volume concentration*, if the specific surface area (→ *surface area, specific*) of the involved pigments and fillers (extenders) is known. However, in practice rather empirical criteria, concerning the application properties of the corresponding coating material, are taken into consideration in formulating pigmentations. For example in the formulation of a primer surfacer, which has to achieve a high corrosion protection, one should try to get an optimal → *barrier effect* using platelet shaped fillers (extenders) like talc. A similarly required filling power can be reached by the addition of fine, compact extender particles, for example of barium sulphate, which fill the relatively large spaces between the coarse talc particles without destabilizing the barrier thus formed.

paint and coating drying: Solidification process of a coating film, due to the evaporation of the solvents. Often, the more precise term "physical drying" is also used. In this regard, some technologically important intermediate stages can be differentiated (→ *initial drying* and *dust-free drying*). Unfortunately, some times processes of the coating technology are also designated as drying (for example oxidative drying), while in reality they are chemical cross-linking reactions. In these cases it would be more correct to use the term "hardening".

paint remover: It serves to remove old coating layers. As a rule, it contains solvents with a high swelling power, often supported by alkali.

paint system: Coating system, consisting of different coating materials, which are produced for craftsmen application. Here, the single layers fulfil specific functions and have to be coordinated with each other optimally.

parameters: Factors of influence in a process, on which depend the process results and their reproducibility. In the coating technology the "dispersing parameters" are particularly important. These are partly unchangeable, because they only depend on the design of the dispersing unit, therefore these unchangeable ones are also called machine parameters (shape and size of the grinding vessel, number and shape of the rotary discs etc.). The changeable parameters like flow rate of the mill base, temperature, speed of the rotor, bead fill level in the grinding vessel have to be optimally worked out for each coating material by the process engineers and kept constant by means of control technology.

parquet sealing coat: Transparent, closed pore coating of wood parquet floors (→ *open pore*). Parquet sealing coats are steadily exposed to wearing by abrasion, thus they have to show high mechanical resistance. In industrial parquet coating, radiation curing clear coats on the base of unsaturated polyesters or modified acrylics are almost exclusively used. For manual parquet sealing, two-pack polyurethanes are mostly employed but also acid curing coatings on the base of polyester/urea resins are as well.

particle: Limited and separable unit of a solid substance. In the coating technology, this term refers mostly to pigments and fillers (→ *primary particles, aggregates* and *agglomerates*). In some cases it is also used for film formers (for example in dispersions or colloidal solutions) and additives (for example flatting agents, waxes etc.).

particle shape: It influences, to a considerable extent, the application properties of fillers (extenders) and pigments. Thus, platelet shaped particles like those of → *talc, kaolin* and *micaceous iron oxide* have a pronounced barrier effect and provoke at the same time, even in relatively little amounts, increase in viscosity as well as gloss reduction in coating materials, while particles with a compact shape, like those of barium sulphate and calcium carbonate, cause only a low flatting effect and have a weak influence on rheology. Concerning the → *aluminium bronzes,* the mirror effect depends, among other things, on the uniformity of their platelets. Aluminium platelets with smooth, circular rim (silver-dollar types) show for example a more brilliant effect than those with rugged rim. As per needle shaped iron oxide yellow pigments, their colour shade depends to a large extent on the proportion between the thickness and the length of the needles. Also the pronounced clear grinding sensitivity of the iron oxide yellows has to be attributed to the needle shape. As a last example, the different colours of phthalocyanine pigments with different crystal modifications should be noted. Although they have the same chemical structure, their different light absorption behaviour along the different crystal axes (anisotropy) generates diverging colour shades.

particle size: Like the particle shape, also the particle size has a relevant influence on the application properties of fillers (extenders) and pigments; but in addition to that it plays an important role relative to the behaviour of film former solutions and dispersions. Fillers are mostly employed in different particle size ranges, depending on the technological properties they have to influence the coating material (→ *particle size distribution*). Thus, in thick film systems, for example heavy duty corrosion protection or road line paints, as a rule only fillers with coarse particles are used, because here the requirements concerning gloss, barrier effect, reinforcement function, abrasion resistance and finally economy (coarse fillers are logically less expensive than fine ones) are completely different from, for example, those for an automotive primer. Concerning the pigments, the particle size influences decisively the scattering behaviour as well as the absorption efficiency of the pigment particles (→ *light scattering* and *light absorption*). So for example, titanium dioxide achieves its optimal scattering power and therefore its maximum whitening and hiding effect only with a particle size about 400nm. Concerning many organic and carbon black pigments, the optimal absorption power develops with particle sizes below 200nm, thus within the transparency range. This is an important target for the use in metallic coatings, however it can be a disadvantage in plain colour coatings because of the insufficient hiding power (→ *transparency*). In the case of binders, the particle size is a measure to define different states of solution (→ *solution, true; solution, colloidal* and *emulsion*), which have major importance relative to the application properties of the corresponding coating materials.

particle size distribution: The powdery components of coating formulations like pigments, fillers (extenders), solid additives (for example high dispersed silicic acid, flatting agents), which are employed as raw materials, but also the powder coatings show a specific particle size distribution. This means that their particles do not have all the same particle size, but the particle sizes vary within defined limits, according to statistical laws. Beside the main range, there are also finer and coarser particle size ranges. The coarser particles can cause, in some cases, coating defects like speck formation and gloss reduction. Too large amounts of fine particles can also provoke undesired changes of properties (for example aluminium bronzes become with increasing amount of fine particles more grey). Screen analysis (sieving) is the simplest but also most time-consuming method for the evaluation of the particle size distribution. More modern, electronically assisted methods, for example "Coulter Counter" or "laser light diffraction", provide quicker and more exact information.

passivating: Formation of corrosion protection layers onto metal substrates by means of chemical reactions. This can take place by the use of phosphate pigments in a → *primer* but it can also be the result of a special pre-treatment of the metal. In this regard, the phosphating of iron and the → *chromating* of aluminium should be mentioned.

pearlescent pigments: → *interference pigments*

pencil hardness: Hardness evaluation method, which utilises pencils with different standardised hardness degrees. These are drawn with their point across the surface of the painted test panel, one after another and in increasing order of hardness, at constant pressure and defined angle. The hardness value is expressed as the grade of the pencil, which has left behind the first visible trace on the coating surface.

pendulum hardness: Better called "pendulum damping according to König" (DIN EN ISO 1522). It is one of the most often used methods of hardness measurement in the coating technology. A pendulum, equipped with two supporting balls, is laid upon the coated surface so that only the balls have a contact with the coating layer. After that, the pendulum is put in oscillation by means of a mechanical device at a defined amplitude. The hardness is expressed as the number of oscillations, which are necessary to reach, as a rule, an amplitude level half the initial value. The softer the film is, the more the pendulum is damped and thus the smaller is the number of the necessary oscillations. The opposite happens in case of increasing hardness.

pentaerithritol: It is the most important tetravalent alcohol ($C(CH_2OH)_4$) for the manufacturing of polyesters. Due to its compact structure, it renders possible a high cross-linking density and therefore the formation of polyester resins with higher hardness and softening temperatures. It is particularly interesting for powder coatings.

peroxides: Chemical compounds, which have two oxygen atoms directly linked with each other. This particular structure causes easy decomposition into reactive radicals, which are used to trigger polymerisation reactions (→*polymerisation*). Besides their employment as initiators in the production of film formers, according to the above mentioned reactions, peroxides are also used in the coating technology as so called "hardeners" for unsaturated polyesters. Here they are added in a stabilised form, as a paste or solution, to polyester putties or surfacers a short time before their application. Since the polymerisation reaction is immediately started and, depending on the system, the →*pot life* may be quite short (for example regarding putties for automotive refinish only 5 to 10 minutes), often a quick working method is required. During the processing of peroxide solutions, attention always needs to be paid because splashes can be very dangerous for the eyes (Put on Your protective goggles!). Because of their oxidising properties, peroxides are not permitted to be stored together with combustible substances and must be constantly cooled during storage. Besides, they never should be mixed with driers containing metals (Danger of explosion!).

phase: System of substances with an uniform state of aggregation (gaseous, liquid or solid).

phase interface: It is the surface area, where two different phases contact each other, for example air and a liquid, a binder solution and the solid surface of a pigment, water and the surface of the droplets of an emulsified resin. At the phase interface of such systems takes effect the →*surface tension* of the involved substances, whose difference is expressed as →*interfacial tension*. The precondition for the stability of a multiphase system is a low interfacial tension, which often can be reached only by the addition of surfactants (→*surface active*).

phase separation: Result of the instability of finely dispersed multiphase systems. The "breaking" of dispersions and, as an only apparent phase separation, the "settling" of pigments in coating materials are particularly important for the coating technology. In the first case, the binder droplets or beads, which swim in the water, flow into each other, because of too strong shear forces or too high temperature and form a liquid or solid phase, which separates from the aqueous phase. In the second case, due to an insufficient thixotropy in the coating material, pigments with a high density tend generally to sedimentation. This is not necessarily negative, if the pigment particles stay still enveloped in a stable binder wrapper and therefore the sediment can be stirred up without speck formation. In contrast to this, the so called "hard cake sedimentation" is a problem, because here the pigment particles have lost totally or partially their binder wrappings and therefore have formed a compact phase on the ground, which can hardly be stirred up in a proper way.

phenolic resins: They were the first entirely synthetic resins and, at the time of the oil varnishes, they displaced the natural copals (tree resins) because of their higher resistance. These synthetic resins, according to the chosen proportion of the reaction components, can be either only physically drying (novolaks) or chemically curing (resols). Resols are generally hardened by stoving in combination with partners like polyesters, alkyd resins and above all epoxy resins (→*gold varnishes*). But the possibility of curing at room temperature also exists, by means of acids as catalysts, which however takes much more time than hardening by stoving. Due to their pronounced hardness and brittleness, the phenolic resins always need to be combined with plasticising partners. In these combinations, besides the exceptional mechanical properties, they also provide the films with outstanding chemical and water resistance. A disadvantage of the phenolic resins is their high tendency to darken. Therefore, use in bright pigmented systems has to be excluded. Besides the interior can coating, the main fields of employment for the phenolic resins are highly chemically resistant coatings, electrical insulation coatings and acid curing primers.

phosphate pigments: They are the most important anticorrosive pigments for the industrial corrosion protection (→*corrosion protection, OEM, refinish, industrial etc.*) sector. They were introduced into the market in the 1970s to replace the carcinogenic zinc chromate and zinc tetraoxi-chromate and ever since they have dominated this segment of application. Besides the straight zinc phosphate, combined pigments are also used, through which by the addition of zinc oxide, other phosphates or organic zinc inhibitors the weak initial corrosion protection has been improved. The corrosion protection effect is based upon the formation of passivating layers (→ *passivating*).

phosphate plasticisers: Easily gelling plasticisers (→ *gelling power*) with fire-resistance properties. The best known representative of this group is the tricresyl phosphate. They are employed in sectors where fire-resistance properties are desired but no high chemical stability is required.

phosphating: Pre-treatment method, preferably for iron and steel substrates. It is also called phosphatising. It has been for a long time a permanent element of the process chain in the automotive production, but it is also widely employed in the coil coating.. The main goals of phosphating are on the one hand the formation of passivating layers (→ *passivating*), which contributes substantially to corrosion protection, and on the other hand an enlargement of the contact surface to the coating, which improves adhesion. There are two different methods: the "iron phosphating" and the "zinc phosphating". In the first process, a very thin iron phosphate layer is formed by the reaction of phosphoric acid with the substrate in an alkaline medium. In the case of zinc phosphating, a thicker film of crystalline zinc phosphate is formed, by precipitation from a solution containing zinc phosphate, which shows a marked surface roughness and, similar to sanding, produces better adhesion.

phosphatising: → *phosphating*

phosphoric acid: It is the most important inorganic acid for the coating technology. It is employed in the passivating of metal substrates (→ *phosphating*) but also as hardener component for two-pack primers. In wash primers it causes, due to a complex chemical reaction with the metal substrate and the main film former component (→ *polyvinyl butyrale*), outstanding good adhesion even on difficult metal substrates like aluminium and zinc.

photochemical degradation: Premature degradation of polymers, which are used as binders in coating materials, due to the influence of ultraviolet radiation and the support of photo catalysts. In the coating technology, this phenomenon can be observed above all when non-coated titanium dioxides are employed. The most important white pigment titanium dioxide, even in its rutile modification which is particularly interesting for the coating sector, shows a remarkable photo-catalytic effect. This means, that it is able to absorb ultraviolet radiation and to start a photo-chemical reaction chain, which leads to intensive film former degradation at the interface between film former and pigment. Therefore, the pigment particles at the deteriorated surface lose more and more of their fixing in the binder matrix, until they form a loose layer, which can be wiped like chalk (→ *chalking*). The precondition for the successful employment of titanium dioxide pigments in coating systems is an after-treatment of the pigment particles, which leads to a coating of their surface. Thus, the photochemical degradation process is forestalled.

phthalate plasticisers: They are the most widely employed plasticiser types. Chemically, they are esters of the phthalic acid. They are characterised by good → *gelling power*, high compatibility, colourlessness and low price. A disadvantage is their low chemical resistance. Their main application fields are the plastification of nitrocellulose and polyvinyl chloride and the manufacturing of pigment preparations. The most important representatives of this class are → *dibutyl phthalate* and → *dioctyl phthalate*.

phthalic acid: It is the most important dicarboxylic acid (bivalent organic acid) for the synthesis of all variants of polyesters, which are employed as film formers in the coating technology. It can appear in three different structures (isomers), that is ortho-, para- and meta-phthalic acid. The most often used variant is the ortho-phthalic acid, which is almost exclusively employed in its dehydrated form as phthalic anhydride.

phthalocyanine blue pigments: → *copper phthalocyanines*

phthalocyanine green pigments: → *copper phthalocyanines*

pH-value: Measure for the acid properties of an aqueous solution. For practical applications it is important to know that a pH-value of 7 expresses a state of neutrality, while values below 7 correspond to acid and those above 7 to alkaline solutions. In the coating technology the pH-value plays an important role, above all concerning the stability of water-borne coatings. Undesirable changes of the pH-value can make reversible the mechanism of water solubility and so destroy the stability of water-borne dispersions or emulsions (→ *water-borne coatings*).

physical drying: Drying process of a coating material, which develops without any chemical transformation of the film former. In the physical drying, the film forming is only caused by solvent evaporation. Since here no chemical cross-linking of the binder takes place, the film former can always be dissolved again by the solvents, in which it was dissolved before drying.

physical properties: Properties, which are inherent to existing substances and not connected with their transformation into other substances. Physical properties are for example viscosity, melting point, electrical conductivity, refractive index etc.

physiological harmlessness: Absence of danger potentials for the human health. Obviously physiological harmlessness is an important goal for all coating materials, though in many cases it can be achieved only partially.

pigmentation: Part of the coating formulation, where the kind and amount of the pigments are fixed, which are necessary to achieve the required colour.

pigment content: Important parameter for the design of pigmented coating formulations. In practice it refers to the total amount of all coating components in powder form, that is the pigments and the fillers (extenders). The pigment content depends directly on the → *pigment volume concentration* of a coating material and it influences several technological properties like hiding power and tinting strength, gloss, flow properties, corrosion protection, water vapour permeability and different mechanical properties.

pigment lakes: Products of a special process, whereby soluble organic dyestuffs are turned into almost insoluble organic pigments. In the coating technology this method is only used for red azo-dyestuffs, from which, by normal molecule increasing processes, only azo-pigments with a pronounced sensitivity against solvents can be derived (→ *bleeding*). During the lake process, the carboxyl groups, which are contained in the azodyestuff, are transformed into a salt by reaction with the base of a bivalent metal, as a rule calcium or barium. Thereby the molecular weight of the dyestuff is more than doubled and, due to that, its solubility is relevantly reduced. The so obtained "pigment lakes" show a sufficient overcoatability. However, due to their salt character they are not weather resistant and therefore practically not suitable to outdoor applications.

pigment paste: → *tinting base*

pigment preparation: Liquid, pasty or solid material, in which a pigment is embedded in finely dispersed form, often even free from agglomerates (→ *agglomerates*). Pigment preparations make possible the incorporation of pigments into coatings without expensive dispersing processes, since, as a rule, they only have to be stirred. Emulsifiers (for example in master batches for dispersion paints) or binders with wetting agents can serve as the liquid phase in the production of pigment preparations. As per the solid phase, several solid resins can be taken into consideration possibly with the addition of plasticisers. The preparations are normally manufactured by the pigment producers using special techniques, so that no agglomeration can take place. Worth mentioning is for example the flushing method, where in the aqueous pigment cake, directly after the pigment precipitation, a wetting conversion is provoked by means of special wetting agents and oil modified resins, so that the water is expelled from the cake and replaced by the oil phase. A main disadvantage of the pigment preparations is that through them extraneous binders, emulsifiers and plasticisers are introduced into the coating formulation.

pigment properties: This term regards the properties of pigments in their application as colorants. These properties are in the first instance the → *hiding power* and the → *tinting strength*, the → *colour location*, in addition to the dispersing properties and the resistance against light, weather and chemical effects.

pigments: According to DIN 55 943, pigments are substances, which consist of particles, are insoluble in the application medium and can be employed either as colorants, that is as colouring materials, or because of their anticorrosive or magnetic properties. However, if the term "pigment" is used without supplementary words, in the practice only colorants are meant, which develop their optical effects according to the mechanisms of → *light absorption* and → *light scattering*. Pigments whose appearance depends on mirror effects or light interference phenomena are not simply called "pigments" but "effect pigments", those with corrosion protection properties are called "anticorrosive pigments", and there are also "conductive pigments", "daylight fluorescent pigments" etc.

pigment surface: → *surface area, specific*

pigment testing: In this category fall different possibilities of quality evaluation for pigment supplies. However, in the first place is the testing of colour matching, which, due to the different fields of application, can be carried out according to quite different criteria. The choice ranges from the simple visual comparison of the pigment powders to the quick dispersion in standardised binders, by means of paint shakers or mini bead mills and subsequent manual application with a doctor blade on contrast paper board, to end with the production of a coating in its original formulation and its application by means of a spraying automate which is compared to standardised material and with subsequent colorimetric evaluation.

pigment volume concentration (PVC): It expresses the ratio of the volume of the pigments and fillers (extenders) to the total volume of the dried film (non-volatile matter) in percentage. The PVC is also a measure for the filling rate of the pigment and filler particles in a layer and therefore it influences several technological properties of the dried coating film. With increasing PVC-values, for example hiding power, filling power but also porosity of the layer increase too (→ *critical pigment volume concentration/CPVC*), while gloss and levelling decrease. To calculate the PVC, the density of the involved pigments and fillers as well as of the solid binders must be known, in order to derive the volume ratio from the mass ratio.

pigment wetting: → *wetting*

pilot plant: Term for production plants, where, in a smaller but nonetheless industrial setting the transfer of newly developed products from the laboratory to the pertinent production conditions is tested.

pinholes: They can be caused by little air inclusions, often also contained in the coating in the form of micro foam, which during the drying process, because of their little size, rise in the liquid coating material only very slowly. On their way to the coating surface they leave behind channels, which cannot be closed any more by the coating material, because in the meantime this has become highly viscous, due to the solvent evaporation. Other causes for pinholes can be → *blisters*, which are not inclined to levelling, or a too high → *thixotropy* of the coating material. Beyond such negative effects on the coating appearance, pinholes can also contribute to the decrease in the corrosion protection properties of a layer. Relating to this, micro pinholes, which externally are hardly visible, are particularly disagreeable.

plain colour coating: Application of a top coat, which contains no effect pigments like → *aluminium bronzes* or → *interference pigments* but only "normal" pigments, whose colour formation mechanisms are exclusively the → *light scattering* and the → *light absorption*. As a rule, the plain colour coating is based upon one-coat systems. However, in the automotive coating the trend is towards two-coat systems, where a plain base coat is protected by a clear coat. The advantages of these two-coat systems are better gloss retention, more efficient mechanical protection and the possibility, by addition of light stabilisers, to improve colour stability.

plasma treatment: Highly efficient but expensive variant of the pretreatment of plastic surfaces before application. In a vacuum chamber, the workpiece of plastic is exposed for a few minute to the effect of a plasma, that is a gas containing electrically positively and negatively charged ions besides non-charged atoms and molecules. Thus, the nonpolar plastic surface, which would hardly be able to develop sufficient adhesion forces to the coating layer (→ *adhesion*), is provided with electrically charged points. At the same time, an efficient cleaning of the plastic substrate takes place. This way, coating adhesion is considerably improved.

plasticisers: Substances with low viscosity, consisting mostly of single molecules, which are capable of softening certain brittle, physically drying polymers for a long time. Precondition for this is stable anchoring of the plasticiser molecules onto the polymer molecules by means of physical forces of attraction, according to a mechanism very similar to the solvation. This process leads, at high temperatures or with support of suitable solvents, to a more or less pronounced gel formation (→*solvation* and *gelling power*). The stability of the anchoring, and therefore the migration resistance of the plasticiser, depends to a high degree on its gelling power. On the other hand, the gelling power depends on the right adaptation of the polarity of the plasticiser to that of the polymer molecules. According to their fields of employment, there are many different plasticiser types. The most widely used plasticiser group is that of the →*phthalate plasticisers*, because they combine all-round application properties with a low price.If the combination of a good gelling power for polar polymers (for example nitrocellulose) with flame-retardant properties is required, than the →*phosphate plasticisers* are the best choice. High fire resistance connected with chemical resistance can be obtained by employment of →*chlorinated paraffins*. At last, one can reach the maximum chemical resistance by the addition of diphenoxyethylformal. As an exception it is worth mentioning the application of the physiologically harmless esters of the citric acid as plasticisers in the food and cosmetics sectors (coatings for food packaging, nail-varnishes and lacquers for make-up pencils).

plastics, coatings for: →*coatings for plastics*

plastification: Softening of resins, which, due to their chemical structure, employed as film formers, would generate brittle films. The term refers to the "internal" as well as to the "external" plastification. In the case of the internal plastification, softening molecule segments are directly installed into the structure of the film former, during its synthesis. Examples are, the →*oil boiling* of phenolic resins and the production of alkyd resins. In the external plastification, the softening of the film former takes place by the subsequent addition of →*plasticisers*. Main examples of this are the plastification of nitrocellulose or polyvinyl chloride with plasticisers as well as with soft resins (→*nitrocellulose combination lacquers* and →*plastisols*). The main advantage of the external plastification is the possibility to vary the quantities of plasticiser as one likes. Conversely, the absence of a chemical connection in the system is disadvantageous, because under unfavourable conditions it may lead to a plasticiser migration and therefore to a subsequent embrittlement in the film.

plastification, external: → *plastification*

plastification, internal: → *plastification*

plastisols: Special dispersions of polyvinyl chloride and its copolymers in plasticisers (→ *organosols*). The film forming takes place only at temperatures of about 130 °C in the "gelling drier". Here the plasticiser gels the polymer and forms with it a homogeneous phase, which solidifies by cooling. Plastisols are widely employed above all for road line paints and underbody sealing for OEM.

plate-like: It is the particle shape, which, particularly with regard to the fillers (extenders), promotes the creation of a → *barrier effect* and therefore contributes to an increase in corrosion protection. The fillers with the more distinctive plate-like shape are → *talc, mica* and *micaceous iron oxide.*

polarity: It expresses the state of electrical charge of a substance and indirectly it gives information about how strong its physical interaction with other substances is. The different polarity of substances depends on their chemical structure. The more atoms are able to attract the negatively charged electrons of the chemical bonds, in which they are involved, the more asymmetrical becomes the charge distribution in the substance. The consequence is the formation of so called "dipoles", which show respectively one negatively charged pole, rich in electrons, and one positively charged, poor in electrons (→ *dipole*). The more pronounced the tendency to formation of dipoles or ions is, the more polar is the substance. The polarity plays a decisive role regarding the → *adhesion* and the → cohesion of the matter. In the coating field it is particularly important to know the polarity of the raw materials, because this exercises the widest influence concerning all wetting, dissolution and adhesion processes (→ *wetting* and → *dissolving power*).

polishing: Smoothening of surfaces by means of polishing pastes or agents. The polishing of coatings takes place with polishing pastes, which contain very fine fillers (extenders). By rubbing the coating surface with the paste, using a soft cloth, one causes a microscopic sanding effect, which eliminates the finest roughness. Therefore, high gloss is generated.

polyacrylic resins: → *acrylic resins*

polyaddition: Chemical reaction, through which the functional groups of two substances are cross-linked with each other by the shifting of atoms or atom groups and without any separation of products. The polyaddition is the classical cross-linking reaction for the two-pack systems, like for example isocyanate based acrylics or amino curing epoxy resins.

polyalcohols: Important raw materials for the synthesis of polyesters and maleic resins; in addition they are modification components for other coating raw materials (colophony, toluylene-diisocyanate etc.). The number of OH-groups in their molecules expresses their valence. Bivalent alcohols (glycols) can only lead to linear polyester molecules. Only with tri- and tetravalent types it is possible to generate three-dimensionally cross-linked structures. Examples of bivalent alcohols are ethylene glycol, butanediol and neopentyl glycol. The most important trivalent alcohols are propanetriol (glycerine) and trimethylol propane. Among the tetravalent types, only pentaerythritol plays a role.

polyamides: Products from the reaction between dicarboxylic acids and diamines. They are very important tough and elastic synthetic materials and in this function they are also processed as powders in the →*fluidised bed process*. As film formers for the true coating sector they are only employed in fringe fields (for example electrical insulation coatings), due to their insufficient solubility. However, they are interesting as modifiers for drying alkyd resins, in which they generate a high thixotropy (thixotropic alkyd resins). On the other hand, polyamides play a much more important role in the form of liquid polyamidoamines, that is polyamide chains with two final amino groups, which are used as hardeners for highly elastic two-pack epoxy systems in the heavy duty corrosion protection.

polyamines: Polyamines are used as hardener component for two-pack epoxy systems, instead of polyamidoamines, when harder films with higher saponification and chemical resistance are required. A disadvantage of the polyamines is their higher volatility, which causes disagreeable odours.

polycarboxylic acids: Polyfunctional organic acids, which play a key role in the synthesis of film formers. In the first place, they serve in the production of polyester resins by reacting with polyvalent alcohols (→*polyalcohols*). Most frequently, bivalent types are used, that is those containing two reactive carboxyl groups in each molecule. The most important members of this group are the →*phthalic acid* and the →*maleic acid*. Tri- and tetravalent types like the trimellitic and pyromellitic acids are mostly employed for water soluble resins and powder coating.

polycondensation: Chemical reaction between two polyvalent partners, which, during the cross-linking, causes even small molecules to be separated. Particularly important for the coating technology are the polyesterification (→ *polyester resins*) and the polyetherification. The latter represents the main reaction in the stoving cross-linking processes. According to the different systems, water, alcohols and some time formaldehyde can be generated as separation products.

polyester putties: They are very important for the automotive refinish as well as for the coating of large vehicles, in addition to some sectors of industrial coating. Depending on their application field, they can be equipped with different accelerators and allowed to react by means of different peroxide pastes or solutions, so called hardeners. Depending on the variant, their → *pot life* may be very different. While putties for automotive refinish take for example only a few minutes to harden, putties for large vehicles can exceed one hour.

polyester resins: They represent the quantitatively most important group of film formers. Polyester resins exist in different variants and are processed and cross-linked, according to different technologies. They are produced by → *polycondensation* of polyalcohols and polycarboxylic acids. According to the choice of the acid components, one can produce saturated or unsaturated polyesters, that is such which contain double bonds (→ *polyester resins, unsaturated*). Unmodified saturated polyesters are mostly used for coil coating systems and powder coatings. For information on the application of fatty acid modified polyesters see → *alkyd resins, short oil, medium oil* and *long oil.*

polyester resins, unsaturated: They are produced by the reaction of glycols, phthalic acid and the unsaturated maleic or fumaric acid. The double bonds of the maleic or fumaric acid segments are later used for cross-linking by means of a → *polymerisation*. To regulate the viscosity or the application properties, so called → *reactive thinners* are used, in general styrene, which, during the cross-linking reaction, are involved in the polymerisation and so stay in the hardened film. Thus, the unsaturated polyesters rank among the low emission systems. The polymerisation can be started either by means of organic → *peroxides* or of energetic radiation (→ *radiation curing*). Unsaturated polyesters have major importance in the field of curtain coating of furniture parts (→ *furniture lacquers*). Also, they are employed to a large extent in the production of → *polyester putties* and dripping resins to impregnate winding wires of electric motors.

polyisocyanates: Hardener components for two-pack systems, whose crosslinking mechanism leads to the formation of polyurethanes. The classic cross-linking partners of the polyisocyanates are polyesters or polyacrylates with free OH-groups (→ *DD-coating*). Between the OH-groups and the isocyanate-groups (-NCO) the cross-linking takes place already at room temperature by means of an addition reaction (→ *polyaddition*) and leads, after a certain time, to outstandingly resistant layers (→ *polyurethanes/PUR*). Regarding their structures, one differentiates between aromatic and aliphatic polyisocyanates. The aromatic ones react more quickly but, after cross-linking, they are not light fast enough and so, as a rule, they can only be employed indoor (for example parquet floor coatings, machine coatings etc.) or in two-pack industrial surfacers. The aliphatic ones warrant high light fastness and therefore colour stability and are preponderantly used in two-pack systems for automotive refinish, large vehicles and aviation coatings, but also for wood and plastics.

polyisocyanates, blocked: → *blocked isocyanate*

polymer: Common term for all synthetically produced macromolecular compounds of the organic chemistry, with molecular weight, as a rule, above 2000g/mol. Polymers with lower molecular weight are more often called → *oligomers*. According to their molecular structures, one differentiates between linear, branched and cross-linked polymers. The mechanism of the reaction, which leads to their formation, can be a polymerisation, a polyaddition or a polycondensation. The corresponding reaction products can also be called polymers, polyadducts and polycondensates. In the coating technology, polymers are almost exclusively employed as film formers, yet in some cases also as additives.

polymer dispersion: → *dispersion*

polymerisation: Chemical reaction, through which the molecules of the initial components (→ *monomers*) are linked with each other by means of their double bonds. Precondition for polymerisation is the preceding splitting of the double bonds into reactive radicals, which unite with each other to new simple bonds. The splitting into radicals can take place by the support of organic peroxides or energetic radiation (UV or electron beam curing). If only one monomer is used for polymerisation, the process is called "homopolymerisation". When different monomers are employed, the reaction is called → *copolymerisation*. In the modern use of the chemical scientific vocabulary, the term "polymerisation" expresses also the other reactions, which are connected with chain prolongation, like the polycondensation and the polyaddition.

polysiloxanes: Chemical term for all silicone compounds. It means that in the molecules of this substances oxygen and silicone atoms are directly linked with each other. To the polysiloxanes belong linear structured compounds as well as cross-linked ones. Products with relatively short linear chains are also known as → *silicone oils* and play an important role in the coating technology such as levelling, anti-blocking and anti-graffiti agents. Higher molecular silicone oils are employed as effect agents in → *hammer finish paints.* Semi-solid, linear and loosely cross-linked products are used in engineering as silicone grease or silicone rubber. At last, cross-linked polysiloxanes are widely employed as → *silicone resins* in weather resistant and highly heat-resistant paints and coatings.

polystyrene: As homo-polymer (→ *polymerisation*) it plays no role in the coating technology, because it is too brittle and sensitive to solvents. On the other hand, styrene is in many cases used for copolymerisation, in order to improve particular properties of the coating materials. Its main applications are the copolymerisation with unsaturated polyesters (→ *polyester resins, unsaturated*), in which styrene is employed as reactive thinner, and with polyacrylates for exterior wall paints, in which the styrene amount reduces the saponifiability by the application on alkaline substrates like cement and concrete (→ *wall paints, exterior*).

polyurethanes (PUR): Chemical compounds, generated by the reaction between polyols (polymers, containing OH-groups) and a → *polyisocyanates.* Polyurethanes are characterised by outstanding mechanical as well as weather and chemical resistance and, if they are derived from aliphatic polyisocyanates, also by light fastness. Therefore, they play a dominant role in the automotive refinish sector, but recently in OEM applications as well. They are also employed in industrial and in furniture coating, if their relatively high price is acceptable.

polyvinyl acetate: Important polymer for the formulation of dispersion paints, which are used preponderantly for indoor applications. It is mostly produced by copolymerisation of vinyl acetate with other softening monomers, like for example vinyl isobutyl ether or vinyl propionate, because as homo-polymer it is normally too brittle and it shows a too high → *minimum film forming temperature.* But it is also possible to subsequently mix plasticisers (→ *plastification*) or dispersions of the softer polyvinyl propionate with the straight polyvinyl acetate dispersion. Since polyvinyl acetate can be saponified relatively easily by alkali, it serves also to produce → *polyvinyl butyrale.*

polyvinyl alcohol: Since polyvinyl alcohol cannot be synthesised by a direct polymerisation, because the monomer "vinyl alcohol" is not stable and turns spontaneously into acetaldehyde, it is produced indirectly by the saponification of → *polyvinyl acetate.* As a raw material for the coating technology, polyvinyl alcohol has only small fields of application as protective colloid for dispersion paints and as binder for special paper coatings. However, its importance as raw material for manufacturing of →*polyvinyl butyrale* is considerably higher.

polyvinyl butyral (PVB): It is produced from polyvinyl alcohol and butyraldehyde. PVB is the main film former in →*primers* for metal substrates. By reaction with the phosphoric acid a chemical bond is formed with the metal substrate (mostly aluminium), which provides for outstanding adhesion. Another employment of PVB is its use as plasticiser in stoving systems, for example →*gold varnishes,* which also causes adhesion improvement.

polyvinyl chloride (PVC): In the coating technology straight PVC is only used in the form of →*plastisols* and →*organosols,* due to its poor solubility. The employment of solutions is only possible if the monomer vinyl chloride is polymerised with other monomers like vinyl ethers and esters, which improve the solubility as well as the compatibility of PVC with other coating components. The most important technological properties of the straight PVC for coating applications are its outstanding chemical and water resistance as well as its mechanical resistance. In some applications, its incombustible nature is also important. Conversely, its light and heat stability are poor, because, by effect of UV radiation or of temperatures above 80 °C, a separation of chlorhydric acid takes place. Straight PVC has always to be softened by the addition of plasticisers, because otherwise it would generate brittle films with poor adhesion. The main fields of application for PVCplastisols and organosols are the underbody sealing of cars, road line paints and coil coating. The PVC-copolymers are used in one-coat systems for bare metal substrates, marine paints and floor varnishes and for the textile fire protection.

polyvinyl propionate: Like →*polyvinyl acetate,* and often in combinations with it, polyvinyl propionate is predominantly used for formulations of interior wall paints based upon dispersions. Unlike polyvinyl acetate, polyvinyl propionate is too soft for some applications and, as a rule, it has to be made harder by copolymerisation with esters of the acrylic and methacrylic acid.

porosity: Property, which can be considered a defect or, conversely, a necessary technological characteristic of a coating material, depending on its purpose and field of application. Porosity has to be evaluated as a coating defect, if it appears in the form of pinholes, blisters (→ *blistering*) or undesired air inclusions in the film after application. As a result of these appearances, the permeability of the film against damaging agents like water, salt solutions and corrosion promoting acids increases, so that the protection of the substrate is not insured any longer. In other cases, for example in highly pigmented dispersion paints or in ventilation surfacers for wood, a certain porosity is indispensable in order to permit sufficient water vapour permeability ("breathing").

porosity test: It take place by application of a sufficiently high voltage between a movable electrode, which can be conducted across the film surface and the coated metal substrate. Where pores are present, the insulating effect of the film is not sufficient any more to prevent an electrical breakdown. The breakdown is signalled by the measuring device in the form of a little flash.

pot life: It is the period of time, in which a two-pack system, after mixing of both components, can be processed without quality defects. For the evaluation of the pot life, as a rule, the increase in the viscosity of the just mixed coating material is taken as a basis. To obtain comparable results, it is important to start the evaluation always with the same viscosity value, the same specimen amount and the same curing ratio. Officially the pot life ends, when the viscosity value has doubled. But it is also possible to carry out the evaluation according to individual processing properties. This is necessary if a viscosity evaluation is not feasible as in the case of polyester putties. Also in the case of water-borne systems the end of the pot life, considered as practical processing time, cannot be evaluated by mere doubling of the viscosity value.

powder coating process: It includes all variants of the powder application. These are the →*fluidised bed process,* the electrostatic powder coating (→*EPS-coating*) and the powder slurry application. The latter takes place in the same way as other liquid coatings but it is restricted to few application fields.

powder coating production: The technology of powder coating production diverges considerably from that of liquid coatings, in terms of both the process chain as well as the employed devices. The raw materials for powder coatings are solid at room temperature, because they are employed neither in the form of solutions or of dispersions. However, to disperse the pigments and fillers (extenders) as well as to achieve a sufficient homogenisation of the raw material blend, a liquid phase is temporarily required. This state is generated by short-term melting of the solid binder. Another difference, in comparison to liquid coatings, is that the formulations of powder coatings are not split into mill base production and let down (→*coating production*). All raw materials of the powder coating formulation are subjected to the complete process of the coating production all together, as a mixture. After the weighing of the raw materials, that is the solid binder inclusive of the hardener component, the pigments and fillers (extenders) and, if necessary, solid additives (for example wetting, degassing and flatting agents), an intensive pre-mixing in special tumble mixers takes place. After that, the raw material mixture is lead to the extruder. This is a cylindrical "hot kneader", which works according to the same principle as a meat mincer. It is equipped with a heating jacket, which permits to reach different extrusion temperatures. Yet as a rule the working temperature is only slightly above the melting temperature of the film former, about 100 °C, in order to avoid the start of a premature reaction with the hardener. The extruder can dispose of one or two screws, which rotate and so procure the transport as well as the intensive dispersion or homogenisation of the raw material mixture, which is steadily dosed into its cylindrical runner by a feed hopper. The paste so produced, while leaving the extruder, is pressed between two cooling cylinders and transformed into a strip, which by further cooling becomes solid. After that, the solid material is broken into little chips, which are subsequently ground to a fine powder by means of an impact mill. Than sieving takes place to separate the too coarse and some time also the too fine particle ranges, which could provoke speck formation and flow interference. The powder coating so produced is now ready for processing and can be electrostatically sprayed (→ *EPS-coating*).

powder coatings: They are the coating systems with the lowest emission, because they do not contain solvents at all. Moreover, they can be processed practically without any loss of material, because in the fluidised bed process no overspray is generated and in the → *EPS-coating* the overspray is generally gathered and led back to the application process. Further advantages of powder coatings are safe transport and storage, due to the absence of combustible solvents. Besides, most powder coating wastes can be handled like house rubbish and do not require the expensive disposal by special refuse incineration, which is prescribed for solvent-borne coatings. Therefore, powder coatings have become more and more important in recent years and they are increasingly conquering new application fields. In addition to the classic powder coating types, which are stoved at relatively high temperatures from 160 up to 200 °C, there are now also powder coatings for the automotive application (OEM), which need only 140 °C for curing; while so called "low melting powders" are being developed, which are even suitable for the plastic coating. By a combination of IR radiation which provides well distributed and careful heating of the powder layer, and of UV radiation which causes a curing of the melted powder coating in a few seconds, it is already possible to coat with powder substrates susceptible to temperature like medium-density fibre boards. Notwithstanding these numerous advantages powder coatings show also disadvantages. Their production is relatively complicated (→ *powder coating production*) and it offers no possibility to match colours. That is also the reason why, up to now, in the automotive industry only clear coat powders could succeed. A further disadvantage is the high minimum film thickness, which has to be applied to get a satisfying levelling (as a rule not below 60µm). Also, the levelling of most powder coatings, with the exception of the acrylic powder for OEM, is rather bad. Finally, the electrostatic application requires expensive devices and high safety standards because of dust explosion danger. Powder coatings can be subdivided into thermoplastic, that is non curing powders for the →*fluidised bed process*, and into thermoset powder for the → *EPS-coating*. The most important types for the fluidised bed process are polyamides powders, which are for example used to coat interior parts of dish washers and refrigerators. The thermoset powders are subdivided into straight, acid polyester powders for outdoor applications (curing with epoxy hardeners or hydroxyalkyl amides), straight epoxy resins for indoor applications with high resistance properties (curing with dicyandiamide) and finally so called "hybrid powders", which in addition to a relatively low price show also good average properties. The latter consist of an acid polyester and an epoxy resin, which cure each other. Acrylic powders play a special role in the clear coat application of the automotive coating (OEM), which show extremely good levelling but also excellent light fastness and weather resistance. The so called "powder slurries", which consist of suspension of powder coatings in water and are generally processed like other liquid coatings, have only a limited use.

powder spraying, electrostatic: → *EPS-coating*

pre-dispersion: Process stage of the coating manufacturing, which normally takes place before the dispersion in a bead mill. The pre-dispersion is mostly carried out by means of a dissolver, after the production of the → *mill base*. For this purpose, the speed of the dissolver is increased up to 18 to 21 m/s; in the case of water-borne coatings as a rule until 10m/s. For coating materials, whose main dispersion process takes place by means of dissolvers a pre-dispersion stage is not necessary. The goal of a good pre-dispersion process is to save time and energy in the bead mill.

pre-mixing: First station in the manufacturing chain for the production of pigmented coating materials. Here, all pigments and fillers (extenders) belonging to the coating formulation are first stirred into a prepared binder solution by means of a slow-running dissolver and by the addition of wetting agents. The so formed "mill base" is subsequently thoroughly pre-dispersed by a dissolver at high speed, before going to the next manufacturing station, that is the → *dispersion process.*

preservative impregnation for wood: Treatment method for wood parts, which are in contact with the soil (pergolas, gazebos, wood fences etc.) and therefore have to be particularly protected against biological degradation. The impregnation takes place with water-borne agents and in special vacuum kettles, where the air is artificially removed from the wood pores, in order to permit a deep penetration by the impregnating agent. The impregnating agents contain → *biozides,* which insure effective preservation of the wood for years.

pre-treatment methods: → *pre-treatment of substrates*

pre-treatment of metals: Concerning the pre-treatment processes, as a rule one has to distinguish between iron- or steel-materials and non-iron metals. All metal substrates have to be cleaned from any external contaminating layers, which could influence negatively their coating adhesion. Remains of silicone, preserving oils, drawing lubricants and hand sweat are normally removed by wiping with clothes and solvents. In industrial processes, hot aqueous surfactant solutions are mainly employed. The second step of pretreatment concerns the removal of more or less well adhering metal oxide layers. These are rust and scale for iron and very hard and smooth layers of aluminium oxide for aluminium. Rust and scale can be removed mechanically by sanding or by using wire brushes, and more efficiently by sand blasting or flame cleaning. Aluminium substrates are normally only sanded. If such layers have to be removed completely, it is preferable to employ metal pickling processes. These refer to chemical dissolution of the oxide layers by means of acids or alkali compounds. For iron, acids are preferred (above all phosphoric acid), while for aluminium alkali compounds are generally used. Another possibility to improve coating adhesion (above all of interest to the industrial sector), is to form so called conversion coatings by the chemical transformation of the metal surface. In such case, the zinc and iron phosphating for iron and the chromating for aluminium should be mentioned. At the boundary between pre-treatment and coating is situated the use of "wash primers", which also serve to improve adhesion and are practically indispensable for the manual spraying application on aluminium and galvanised steel substrates.

pre-treatment of substrates: Each substrate needs careful pre-treatment before coating. Concerning the pre-treatment of metal substrates

→ *pre-treatment of metals*. Pre-treatment of plastics requires similar first steps, that is cleaning with surfactant solutions or solvents. The use of solvents must be carried out very carefully, because thermoplastics are most sensitive against solvents (→ *coating of plastics*). Also sanding for adhesion improvement is usual. Anti-static agents or after-treatment with ionised air help against the electrostatic charge of the plastic. However, the main goal of pre-treatment of plastics is to increase the polarity of the plastic surface, which is generally too low to permit sufficient adhesion. Thus, special techniques are available, like → *blazing of plastic surfaces*, the → *corona discharge* and the → *plasma treatment*. In the case of wood substrates, pretreatment is restricted to sanding off the wood fibres, which have erected vertically, or to smoothing the surface by sanding techniques (→ *sanding*). Some times also the impregnation of wood with preservatives and the → *staining* of wood rank among the pre-treatment measures. The main purpose of the pre-treatment of mineral substrates is the consolidation of the substrate to avoid that a chalky surface diminish the adhesion of the other layers. Pre-treatment is done by means of different impregnating agents and → *primers*. In addition, in the case of visible concrete or bricks a → *fluosilicate treatment* can avoid the efflorescence of salts.

pre-treatment of surfaces: → *pre-treatment of substrates*

prevention of air pollution: → *immission protection*

primary particles: They are the smallest particles of a pigment, which show the typical pigment properties in their entirety (→ *light absorption* and → *light scattering*). Primary particles are mostly single crystals, however some time they can also appear in an amorphous form (for example carbon blacks and high dispersed silicic acids). After manufacturing of the pigments, as a rule, primary particles do not exist in a free state but in an agglomerated form (→ *agglomerates*), together with the aggregates (→ *aggregates*). The goal of the dispersion process in the coating production is the complete breaking down of the agglomerates, in order to set free as many aggregates and primary particles as possible, because only in this manner pigment properties like hiding power, tinting strength and colour purity can be fully utilised.

primers: Common term for coating materials, which are applied before the topcoat and which have to fulfil different duties depending on the substrate properties. In the case of mineral substrates for example, primers often also called primer-sealers, are used to consolidated porous building substrates and to regulate water vapour permeability. Concerning wood, primers have the function to decrease its absorbency and some time also to form a sealing layer against migration of wood extracts. Finally, in the case of iron and steel, they have the outstanding function to insure corrosion protection. The anticorrosive primers economically rank among the most important coating materials, because they contribute decisively to maintain durable industrial and transport infrastructure. They are generally equipped with →*anticorrosive pigments* like zinc phosphate, zinc dust, organic corrosion inhibitors but also with platelet shaped fillers (extenders) like talc, china clay, micaceous iron oxide (→*barrier effect*). There is also a group of special primers, which have the function to improve the adhesion quality of problematic metal substrates, like aluminium and its alloys or of galvanised steel and plastics. In the case of metals, they are often also called "wash primers" and are subdivided into one-component and two-component primers. Their main film former is →*polyvinyl butyral*, often in combination with smaller amounts of phenolic or epoxy resins in order to yield better water resistance, and they contain also phosphoric acid which generates chemical bonds with the metal substrate. These bonds are responsible for the outstanding adhesion of the primer layers. As anticorrosive pigment, now practically only →*zinc phosphate* is employed. The formerly predominant use of →*zinc tetraoxychromate* is currently limited to a few special cases. In the one-component primers the phosphoric acid is already contained in the base material, while in the two-component types the acid is added to the base material as "hardener solution" a short time before application. The primers for plastics are, as a rule, not pigmented and they provoke the adhesion only by means of the forces of attraction of their film former as well as of defined effects of their solvent on the plastic surface. The most commonly used film formers are elastic polyurethanes and chlorinated polyolefins.

primer surfacers: Also called →*fillers* or surfacers. Priming materials, which have the task to level out fine irregularities in the substrate before coating with the topcoat, and in the OEM application also to insure good resistance to stone chipping. For this purpose they are equipped with fillers (extenders), which, due to their filling power, are capable to bridge the substrate irregularities (→*extenders and fillers*) and to form a reinforced texture. There are primer surfacers which have to be sanded after application (for example in the automotive refinish), as well as others, in some cases also called wet-on-wet primer surfacers, which after drying are overcoated without sanding (for example in the OEM coating).

production batch: Defined amount of a coating material, which runs through the complete production process, in the discontinuous coating manufacturing. The number and size of the batches, which are necessary to execute a customer order, depend on the availability of vessels, machines and devices in the coating production.

propanol: Polar solvent for water-borne systems and for the production of solutions with high polarity, for example for the electrostatic application. Propanol is also important as cleaning agent for non polar plastic surfaces (no danger of solving). However, its isomer → *isopropanol* is more frequently used.

propellant gas: → *aerosol lacquers*

protective equipment: Important component of safety in the work place, which is prescribed in the regulations for the prevention of accidents as well as in other pertinent safety regulations concerning the handling of hazardous work substances. In the coating sector, protective equipment includes in the first place protective goggles, gloves and shoes; furthermore → *breathing protection* as well as coats for working in the laboratories and overalls for the production workers, made of not combustible fibres. Protective goggles have to save the eyes from splashes of solvents and/or of acids and alkaline substances, during stirring and filling processes. Protective gloves should, above all, impede the absorption of solvents through the skin and so avoid their effect of fat dissolution, in addition to preventing the corrosive or irritant effects of other hazardous substances. Finally, protective shoes fulfil the function of mechanical foot protection, during the handling of containers or other loads. At the same time they provide a sufficient earthing for electrostatic charges, which could lead to spark formation and to the danger of fire and explosion.

protective gloves: → *protective equipment*

protective goggles: → *protective equipment*

protective shoes: → *protective equipment*

pseudoplasticity: Rheological behaviour of non-Newtonian liquids, which already begin to flow by the employment of low shear forces and whose viscosity, by increasing shear rates, decreases (→ *Newtonian liquids*). Liquids, which need a defined minimum shear force to begin to flow (yield point) are called plastic liquids. In the application technology, pseudoplastic liquids are characterised by the following behaviour: before application they show a high viscosity (yoghurt effect); during application, due to the shearing by brush, roller or spray gun, their viscosity decreases significantly; after application the viscosity recovers very quickly, practically without delay, its initial high value. This behaviour is very important for coating materials, which either are applied as thick films, and so run the risk of sagging (dispersion paints, spray putties etc.), or are not permitted to get a good levelling for optical reasons like structured paints and hammer finish paints.

putty: Heavily-filled coating materials for repairing or levelling out large substrate irregularities. The fillers (extenders) contained in the putty serve on the one hand to form a supporting structure, which prevents the coating material from flowing into cracks and dents after application; on the other hand they improve the sanding properties after its curing. Obviously, only soft fillers can be taken into consideration for this purpose. For putty manufacturing, different binders can be employed. The most commonly used are solvent-free ones, because the solvent evaporation would cause volume contraction of the putty and consequently the reproduction of the form of cracks and dents on the coating surface. Besides, since putties are normally applied as thick films, the solvent would have difficulties to evaporate sufficiently quickly. Yet, this does not concern the so called spray and spot putties, which are applied as relatively thin layers to repair already coated surfaces. The latter spot putties are based on nitrocellulose and equipped with quickly evaporating solvents. In the industrial and automotive refinish coating, putties based upon unsaturated polyesters are usually employed (polyester putties). In the wood coating and decorative paint sectors, oil putties are still used. However, due to their slow drying they are more and more displaced by nitrocellulose putties. As a special variant of wood putties can be considered the so called “plastic woods”, which are based on nitrocellulose and contain sawdust as extender. They serve to fill pores and crevices and conserve at the same time the original colour of the wood.

PVC: → *pigment volume concentration*

pycnometer: Device for the evaluation of the density of liquids. It consists of a cylindrical or bottle shaped vessel with a perfectly sealing lid. The lid is so designed, that, during its setting upon the filled vessel, a liquid surplus can escape outwards through a hole in the middle of the lid. Thus, it is guaranteed that the same defined liquid volume is left in the vessel. After weighing the liquid mass, the density can be calculated. If a liquid with known density is used, it is also possible, by means of a difference calculation process, to evaluate also the density of powdery materials like powder coatings or pigments. However, a precondition for that is that the liquid must be capable to fill all interstices between the particles. In the case of very fine pigments, this is only possible by the use of a glass pycnometer with a connecting piece for vacuum application. The methods for density evaluation by the pycnometer are described in the DIN 53 757 and 53 217 or in the DIN ISO 787 and DIN ISO 2811.

Q

quality control: Instrument of the quality assurance and of the quality management processes (→ *quality management*). Quality control attempts to guarantee the compliance of products with the supply specifications of the customers, by continuous testing of the raw materials employed as well as of the finished products in comparison to standardised samples or fixed test values. In the past, expensive quality control measures were considered the only possibility to assure product quality, by recognising in time defective ware at the end of the production chain and separating it before delivering to the customers (rejects). In the modern quality management systems quality control plays a different role, because in it all lines of workers are continuously involved, by systematically improving failure prevention, to make production processes so reliable that rejects are hardly produced. This way, the costs incurred for tests and complaints is reduced to a minimum.

quality management: Common term for the coordinated interaction of all quality assurance measures, which are valid and required in modern customer/supplier relations. Since the 1980s, the minimum requirements for quality management systems (QMS) concerning development, production, release as well as handling of complaints about products and all kinds of related services, have been formulated into the international standards DIN EN ISO 9000 to 9004. In this regard it should be mentioned that the DIN EN ISO 9001 is particularly important for the industrial sector, because it is normally the base for the certification of most industrial companies. These standards lay down the appointment of a quality manager by the company, the production of a QM handbook with a detailed presentation of the company's responsibilities, operational processes and procedures for the purpose of translating the quality standards into practical applications. Furthermore, the company has to pursue continuous improvements in the quality of its products. The evaluation of the comprehensiveness and effectiveness of the adopted measures is carried out by specialist service providers in the form of so called audits (process inspections). If, after a suitable introduction phase, the quality management system is found sufficiently qualified, the company gets a "certificate" from the independent auditor. This has to be confirmed at regular intervals by other audits.

quinacridones: Group of high value organic pigments, which cover the colour range from red to magenta. Due to their very high light fastness, tinting strength and solvent resistance, they are used in plain colours as well as in metallic colours. They rank among the most expensive organic pigments.

q-value: Measure for the position of the PVC (pigment volume concentration) relative to the corresponding CPVC (critical pigment volume concentration). The q-value is given as PVC divided by CPVC, multiplied by 100 and it expresses which percentage of the CPVC the PVC corresponds to. The q-value is particularly used with reference to corrosion protection primers, because there, with the exception of zinc dust primers, it is required to fall clearly below the CPV. Anticorrosive pigment producers mostly recommend q-values about 75 to 80 %. At higher q-values there is the risk of too high water permeability, and therefore of weakening the corrosion protection properties.

R

radiation: Energy transfer in the form of electromagnetic waves or radioactivity. Of particular interest for the coating technology are different ranges of the electromagnetic radiation, whose properties depend on their wavelengths. So visible light may, for example, only contain wavelengths between 380 and 750 nm (→ *colour*). Radiation with shorter waves, which border on 380 nm, is called UV-radiation and is used for radiation curing (→ *UV-curing*). Above 750 nm is the IR-radiation, which is employed for crosslinking by stoving (→ *infrared drying*). Among the radioactive radiation variants, only the electron radiation, that is a pure β-radiation, is interesting for radiation curing (→ *electron beam curing*).

radiation curing: Eco-friendly and energy saving coating hardening process, due to its poor emission and the extremely short curing times (→ *electron beam curing* and *UV-curing*).

radiator enamels: Term of the craftsman vocabulary, used for decorative paints, which are particularly suitable for application by brush on radiators. Like other decorative paints, no matter if solvent or water-borne, they contain oxidative drying alkyd resins, which in spite of the steady heat exposure, must not show a significant tendency towards yellowing. Another important property of these enamels is their outstandingly good levelling even on spots difficult to reach, where by use of an angular radiator brush the coating can only be spread insufficiently.

radical: Particularly reactive atom or part of a molecule, which is generated by the decomposition of a pair of electrons into single electrons. The high reactivity is explained by the fact that each single electron, belonging to an atom or a part of a molecule, tries to form with suitable reaction partners new pairs of electrons. Consequently, new chemical bonds are generated. The formation of radicals is above all important for polymerisation reactions (radical polymerisation). In the coating technology, for example, the curing reaction of unsaturated polyesters (→ *polyester resins, unsaturated*) is started by the decomposition of peroxides into radicals or by the use of radical forming UV-radiation or electron beams. But the oxidative hardening of drying oils in alkyd resins also takes place through a radical mechanism (→ *alkyd resins, long oil*).

RAL register: German colour register, introduced by the long gone "Reichsamt für Allgemeine Lieferbedingungen" (Imperial office for general supply conditions), but which still retains major importance in industrial coatings and architectural paints, particularly in Europe. The colours are identified by a number of four digits, the first of which represents the colour group. So 1 means yellow, 2 orange, 3 red, 4 violet, 5 blue, 6 green, 7 grey, 8 brown and 9 white, black and other achromatic colours. The colours are available in the form of "RAL colour charts" in different gloss degrees.

rapid rotation atomisers: To them belong the "rapid rotation bell" and the "rapid rotation disk", which ideally combine the principle of the electrostatic with the purely mechanical atomisation. In their working methods they hardly differ from each other. Both the bell and the disk have a sharp edge, where the electrical field shows a particularly high concentration of field lines (→ *electrostatic application*). The field lines guide the formed coating droplets to the object, so that practically no → *overspray* is produced and a very good → *throwing power* takes place. The real atomisation is almost purely mechanical, being caused by the very high rotational speeds of up to 60,000rpm. In practice the rapid rotation bells permit an application perpendicularly to the bell plane, that is along the bell axis, and such that it can be directed to the surface of each object, similar to a spray gun. This technology has wide employment in the automotive OEM coating sector. Conversely, the rapid rotation disk can only atomise on the horizontal plane and so it must execute strokes in vertical movements up and down to coat complete surfaces. For this purpose, the parts which have to be coated are conveyed in a so called omega-loop around the spray disc. Rapid rotation discs are employed for the coating of the so called "white goods" in the household appliance industry and of bicycle parts.

raw materials for paints and coatings: Initial products, which serve to manufacture coating materials. The range of the raw materials for the coating technology is extremely wide and variegated. It comprehends "binders" or "film formers", "solvents", "additives", "pigments" and "extenders" (fillers). Details about the importance and the function of single raw materials are obtained at the corresponding terms.

REACH (Registration, Evaluation, Authorisation of Chemicals): This directive on the introduction and use of chemicals was established by the European Parliament in 2007 and seeks to establish a completely new system for evaluating chemical substances. To this end, a special agency (the European Chemicals Agency, ECHA) was founded, which is responsible for the registration of all chemical substance that are manufactured or imported at or above one tonne per year. Where chemicals are manufactured or imported in quantities of ten tonnes or more per year, the users must submit a Chemical Safety Report (CSR) containing detailed information on the environmental and human health hazard properties. Authorisation depends primarily on the data in the CSR and may be contingent on mandatory replacement of hazardous chemicals by less dangerous substances.

reaction: Chemical process, which leads to a material change. Chemical reactions are the basis for the production of most raw materials, which are employed in the coating technology, as well as the key to chemical crosslinking of the film former components, during the curing processes of coatings. A precondition for the latter is that the employed partners contain reactive groups, which are able to link with each other (→ *cross-linking*).

reactive pigment: Pigment, which is capable to react chemically with components of the coating and/or with the substrate. Most of the reactive pigments are used for corrosion protection, like for example → *zinc phosphate* and formerly zinc chromate. But also basic pigments, which can form soaps with acid film former groups like zinc oxide and white lead, belong to this group.

reactive primer: Component of a curtain coating process (sandwich process), to coat wood with unsaturated polyesters, which today is only seldom employed. The reactive primer contains the organic peroxide and it is separately applied as the first layer. Then, the application of the polyester layer follows. At the boundary between the two layers, the peroxide diffuses from the reactive primer into the polyester layer and there it starts the polymerisation reaction (→ *polyester resins, unsaturated*).

reactive thinners: They serve to adjust the viscosity of reactive coating materials and therefore insure their optimal workability. However, unlike the normal solvents, which have to fulfil the same task in coatings, the reactive thinners do not leave the system by evaporation during the drying process, but they participate in the curing reaction and are almost completely incorporated in the solid coating film. Reactive thinners make also possible the formulation of emission-free or low emission systems. The commonly employed reactive thinners are styrene for unsaturated polyesters, glycidyl ethers for two-components epoxy systems and monomer, liquid acrylic acid esters for beam curing systems.

recycling: Modern international and national rules concerning waste disposal, incorporate recycling as a basic target. However in the coating technology recycling processes turn out to be difficult, because of the variety of formulations and colours. An exception are the powder coatings, whose overspray during their application is almost completely collected and recycled. In the case of liquid coatings, a recycling of overspray can only succeed with one-component materials and equality of colours, and it requires complex technologies. More often employed is the recycling of coating coagulate from the sewage water of application devices by means of ultrafiltration. But in most cases coating waste has to be incinerated as special waste. Processes for the thermal separation of certain coating components and their employment for energy recovery are now in the testing stage.

reduction: This term has two different meanings. The first refers to the coating production and is also called "making up". In this case, it means the reduction of the concentration of pigment pastes or → *mill bases* by addition of binder solution. The second one is a chemical term and it means the contrary of → *oxidation*. Referring to iron it would mean the inversion of the rust formation, that is the transformation of rust into bare iron, something that however cannot be achieved by use of anticorrosive techniques. A reducing medium is advantageous for the corrosion protection. The formation of the reducing agent hydrogen in the cathodic electrodeposition coating is one of the reasons why this process, relative to corrosion protection, is clearly superior to the anodic one, in which conversely the oxidising agent oxygen is formed (→ *electrodeposition, anodic* and *cathodic*).

reflection curve: Curve, which shows the amount of reflected visible light, depending on the different wavelengths, when a → *surface colour* is illuminated by a white light of a known composition. The measuring takes place by means of a spectrophotometer, which is equipped with a spectral sensor line. The test panel or object is illuminated by a standardised white light, and the device evaluates how much of each wavelength range of this white light between 380 and 750 nm, that is along the whole visible spectrum (→ *colour*), is reflected by the coloured specimen. Based on the measuring values obtained this way, a reflection curve can be defined in a coordinates system, with the light intensity on the perpendicular axis and the corresponding wavelengths on the horizontal axis. By means of the reflection curve, the tristimulus values of the measured colour can be calculated (→ *colour measurement instruments*). The matching of the measured reflection curves of colour samples is the basis of the theoretical calculation of the best and metamerism-free pigmentation of unknown colours in the modern colour formulation calculation programs (→ *metamerism*).

reflectometer: → *gloss measurement*

refractive index: It is a measure for the deflection of a light ray with defined wavelength, when this, passing through the air, enters a new material. Depending on the "optical density" of this material, the ray can be deviated to the perpendicular or to the contrary direction. Thus, an angle between the ray and the perpendicular is generated (angle of refraction), which is different from the corresponding angle in the air (angle of incidence). Mathematically, the refractive index is the ratio of the sine of the angle of incidence divided by the sine of the angle of refraction. The measurement of the refractive index (→ *refractometer*) serves to test transparent liquids, like solvents and clear coats. But the refractive index gives also important information about the optical behaviour of pigments (→ *light scattering*).

refractometer: Device to measure the → *refractive index* of liquids. In the coating technology it serves, as a rule, for the evaluation of the purity of solvents. Since the refractive index of a substance also depends on the temperature and the wavelength of the employed light, refractometers are usually connected with a thermostat and work, if there are no other indications, with a sodium lamp.

regeneration: In the coating technology this term refers, as a rule, to recycling processes for solvent mixtures, which are produced as solvent waste or dirty cleaning thinner, during the coating manufacturing. The regeneration is mostly carried out by specialised companies, which distil the dirty solvent mixtures and thus recover a part of the solvents in a pure form. The so regenerated solvent mixture has a non-defined composition and is mostly employed by the coating companies as cleaning thinner. The residues of the distillation process which cannot be used are incinerated as special waste.

relative humidity: The percentage that the current humidity value is of the maximum value, which the air can absorb at the given temperature (saturation value). The relative humidity is above all important for the coating application. In the case of solvent-borne systems, it is for example important to pay attention that, during the application, the temperature does not sink to the "dew point". At this temperature, the saturation value would be exceeded, so that, similar to fog formation, water droplets would separate from the air ad fall down onto the wet coating layer with the corresponding negative consequences. Waterborne systems, in the case of high requirements like for example in the automotive industry, have to be applied in conditioned rooms, where the relative humidity value changes only within a defined field, because the drying velocity depends decisively on the moisture content in the air.

release agents: Substances, which are above all employed in the production of plastic components (for example according to the injection moulding process), in order to facilitate their separation from the casting mould. Naturally, traces of the release agents represent a danger for the coating adhesion and have to be removed completely before coating (→ *coating of plastics*).

reproducibility: It is the capability of a process to always yield the same results. Reproducibility has major importance above all for measurement and test methods, because the comparability of product properties depends to a great extent on it. Decisive for reproducibility is the complete elimination of random factors.

resins: Mostly solid, transparent or translucent organic materials, which, similar to glass, do not have an exact melting point but a melting range. Rarely they are colourless. They normally show a more or less marked yellowish colour, which some time tends to reddish brown (→ *phenolic resins*). In the coating technology, this term is used for all kinds of film formers. In the past, natural resins were often employed like dammar, copal or shellac, while today they have no importance as film formers, with the exception of colophony. Meanwhile tailor made, synthetically produced resins, form the base of the binder industry. These are either dissolved in organic solvents and/or in water or they are finely dispersed in water (→ *dispersion*). However, some times they are also used in a solid form (→ *powder coatings*).

resistance: Common term to express the stability of raw materials or of coatings, which contain them, against all kinds of aggressive conditions. In its specific applications, this term must be defined more exactly (alkali resistance, solvent resistance, weather resistance).

resistance, electrical: According to the Ohm`s law, electrical resistance is defined as the voltage divided by the current intensity. In the coating technology film formers which are used have prevailingly a high electrical resistance and are therefore good insulators. Some of them are particularly suitable in the production of → *electrical insulation coatings*. On the other hand, high electrical resistance, that is at the same time low conductivity, is a disadvantage in the case of electrostatic application (→ *electrical conductivity*). A further problem caused by low conductivity is the danger of electrostatic charge, which can lead to sparkle formation and therefore to fire.

resite: Term for a phenolic resin, which has completely cross-linked with itself. Due to their three-dimensional structure, resites are → *thermosets*, that is they are not soluble and fusible any more. Resites can show different degrees of brittleness, depending on their provenience, that is if they have been generated from non-plasticised or plasticised resoles. However, generally they are all very hard and chemical resistant.

resole: Term for a reactive phenolic resin, which is capable of cross-linking. By employment of substituted phenols or by boiling with oils (→ *oil boiling*), resoles can be plasticised (plasticised phenols). Yet the most often used variant of plastification is the combination with other film formers in stoving or acid curing systems. Particularly worth mentioning is the combination with epoxy resins (→ *gold coatings*).

retention: → *solvent retention*

revolutions per minute: It is an important parameter in the coating manufacturing processes for all dispersing units, which work with a rotating shaft or a rotor. In the first place this applies to → *dissolvers* and attrition mills (→ *bead mills*). In the case of the dissolver, the revolutions per minute serve to calculate the peripheral velocity of the disk, which, beside the geometry of the stirring vessel, represents the most important parameter to achieve an optimal dispersion result. The reason for this is that here the dispersing effect on the pigment agglomerates is only provoked by the shear velocity gradient between the single liquid layers (→ *dispersion process*).

rheological additives: Agents, which are employed to generate thixotropy and/or pseudoplasticity in coating materials. The rheological additives play a particular role in the coating technology, above all concerning the application properties of coatings. There are practically no coating formulations not containing at least little amounts of these additives. The reason is that, for the application of coating materials on vertical surfaces, a Newtonian flow behaviour (→*Newtonian liquids*) would be disadvantageous because the necessary viscosity increasing in order to avoid the formation of →*sag* and →*curtaining*, would only be caused by solvent evaporation and therefore would take place too slowly. In order to produce relatively thick wet coating layers by a single spray cycle, and obtain at the same time an optimal levelling, a thixotropic behaviour is ideal, because it permits the coating material to spread sufficiently, before the rapid increasing of viscosity begins (→*thixotropy*). In the case of coating materials, which have to be applied in very thick layers (for example, thick-film surfacers and dispersion paints) or which have to maintain application structures (structured paints, hammer finish paints), no levelling at all is desired. For these coatings →*pseudoplasticity* is the suitable flow behaviour. Rheological agents are often able, depending on their employed amounts, to cause thixotropy as well as pseudoplasticity. Besides they fulfil also the task of →*anti-settling agents* in highly pigmented systems. The rheological agents are normally subdivided into inorganic and organic types. Inorganic are, for example, pyrogenic silicic acids (→*high dispersed silicic acids*) and phyllosilicates like bentonite. Among the organic variants, derivatives of castor oil play a certain role in solvent-borne coatings. For water-borne systems, beside the traditional thickener on the base of cellulose (→*methyl cellulose*), acrylic thickeners and so called associative thickeners are increasingly used.

rheology: Lore of the flow and deformation behaviour of substances. Concerning the coating materials, the different kinds of flow properties have main importance for their handling, predominantly during application. Of importance are above all the Newtonian behaviour (→*Newtonian liquids*), the pseudoplasticity and the →*thixotropy* as well as their converse variants of flow behaviour "dilatancy" and "rheopexy". However, the latter two occur very seldom in coatings.

ricinolic acid: Non drying fatty acid, since it contains only one double bond (→*fatty acids*). It is the only oil component with an OH-group. This gives it a particularly good compatibility to polar film formers, in addition to suitability to separation of a water molecule (→*castor oil, dehydrated*). It is the main component of the castor oil.

ring circuit stability: Aspect to be taken into consideration, during the development of coatings for the industrial application in processes where the coating material supply takes place by means of circulation systems such as those used in the automotive coating (OEM) or the coating of other large objects. During the transport, but above all in valves and pumps, the coating material is subjected to considerable shear forces which, for example, can destabilise film former dispersion or lead to a post-dispersion of pigments or to damage to effect pigments. These negative processes would provoke an unacceptable change of the colour.

road line paints: Paints with high abrasion resistance, outstanding water and salt stability and marked light reflection also by nocturnal illumination. As film formers polyvinyl chloride copolymers are primarily employed, but increasingly also solvent-free two-component systems on the base of isocyanates and, for hot processing, different unsaponifiable thermoplasts. The abrasion resistance is achieved by addition of hard fillers (extenders) like silica flour and feldspar. In order to generate high light reflection at night, special hollow microspheres of glass are mostly incorporated into road line paints.

rod bending test: Also called mandrel flex test. Standardised test of the resistance of coatings against cracks, caused by deformation. There is a variant with cylindrical (EN ISO1519) and one with conical mandrel (EN ISO 6860). In the first case, a coated metal strip is bent around a cylindrical mandrel of steel, with known diameter, by means of an appropriate device. For this test, a complete set of mandrels with different diameters is available. The diameter of the specific mandrel is registered, when the coating shows the first cracks. With the conical mandrel, the coated strip is bent at the same time around different curvatures and subsequently evaluated.

roller: Tool for the manual application of paints and coatings. According to the different properties of the materials to be coated, different types of rollers are employed, for example lambskin roller or foam rubber roller. Unlike the brush, the roller permits the quick and relatively uniform application of coating materials on large surfaces, provided they are even. On the contrary, objects with angles cannot be coated so well only by means of a roller, therefore a brush must be often used too as supporting tool. Together with the brush, the roller represents a cheap and flexible working tool, which in addition does not provoke practically any loss of material. On the other hand, the quality of the coated surface cannot be compared with what can be achieved by spraying application.

roller coating: One of the most important industrial processes for the coating of metal strips (→*coil coating*), flat wood components for the furniture industry, foils and paper. A precondition for the successful employment of this process is that the workpieces to be coated are completely flat. The roller coating permits the application of very exact coating layers between 3 and 100 µm, besides the printing and stamping of all kinds of surface imitation (for example rare wood types). The coating application takes place by means of pairs of rollers. A chromium-plated feed roll, which is steadily supplied with material from a coating trough, transfers the coating onto a rubber-covered applicator roll. The object to be coated is moved under the rotating applicator roll at constant velocity and receives from it a defined coating amount. The film thickness is regulated by the gap between the feed and the applicator roll. By employment of two rolls one after another, also twopack materials can be applied in a two-coat process, so that the hardener and the base material layers are rolled one upon another.

room temperature: It is the temperature, which arises in a room, due to the existing climatic conditions. It is also not influenced by heating or cooling devices. The room temperature appears often in test specifications or data sheets.

rotation viscometers: Devices for the measurement of the flow behaviour of liquids (→*rheology*). Differently from the easy to handle and cheap →*flow cups*, which, strictly speaking, are only permitted to measure Newtonian or lightly thixotropic liquids, the expensive rotation viscometers make it possible to evaluate all kinds of viscosity anomalies, since they are able to measure stress and shear rate in relation to one another. For clarity purposes, it is necessary to explain that scientifically viscosity is defined as stress divided by shear rate. In this context, stress is the force per surface area which is necessary to move two (imaginary) liquid layers against each other with a certain velocity (shear rate). In the case of Newtonian liquids, this ratio is always the same, provided that the temperature is constant, therefore the viscosity of such liquids depends only on the temperature (→*Newtonian liquids*). In the case of thixotropic, pseudoplastic, rheopexe and dilatant (shear thickening) liquids, dependence of the viscosity on the intensity and duration of the shear forces has to be added to the dependence on temperature. In modern rotation viscometers it is possible to fix stress as well as shear rate and shear time. According to the required task, three different measurement setups are possible: plate-plate, plate-cone and cylindrical coaxial measuring system. According to the type of stress evaluation, one differentiates between Searle- and Couette-system. In the Searle-system, the upper-plate, the upper-cone or the interior cylinder rotate, while the measuring cylinder or the under-plate are stationary. Conversely, in the Couette-system the under-plate or the measuring cylinder rotate. Because of the dependence on temperature, rotation viscometers are generally connected with thermostats.

roughness height: → *roughness measurement*

roughness measurement: Today it is almost exclusively carried out by the electromechanical wavescan method or by scanning the surface with laser beams. So profilograms are produced, which can be mathematically evaluated in different ways. Most frequently, the mean roughness height according to DIN 4768 is indicated.

rubber derivatives: → *chlorinated rubber* and *cyclised rubber*

rub out test: Simple method for evaluation of flocculation or floating phenomena in pigmented coatings (→*flocculation and floating*). Thereby, one let the applied coating dry, until it begins to stick. After that one point of the coating film is rubbed with the fingertip, but without reaching the substrate. This way, the flocculates are destroyed and eventually separated pigment particles are mixed with each other, so that the original colour is approximately restored. If a colour difference between the rubbed point and the surrounding coating film becomes visible, which can also be evaluated by colour measurement, than a pigment separation has taken place in the wet film.

rust: Product of the oxidation of iron and steel (in the case of white rust also of zinc), mostly due to weather exposure. Rust forms a brown, porous, flaking layer, which above all consists of hydrated iron oxide. Its poor solidity and insufficient adhesion causes the progression of the corrosion process until the material destruction (→ *corrosion*). Thus, in the coating technology the avoidance of rust formation by means of protective coatings (→ *corrosion protection*, OEM, refinish, industrial and heavy duty) plays a particularly important role from an economic standpoint.

rust prevention primers: → *primers*

rust protection: → *corrosion protection, OEM, refinish, industrial* and *heavy duty*

rust removal: Necessary pre-treatment step before coating of iron and steel, because on already formed rust neither an efficient corrosion protection nor a sufficient adhesion are possible. The rust removal can take place manually, by means of a wire brush and/or sandpaper, or by mechanical techniques like blasting with steel grit, glass, corundum or flame cleaning.

rutile: → *titanium dioxide*

S

safety: → *occupational safety*

safety data sheet: Standardised information about a substance, prescribed by law and with detailed indications regarding labelling in accordance with → *REACH* and → *GHS*, as well as handling during transport and storage, including the designated measures in the case of fire and accidents. It has to be made available by the producer to the user at the latest on the occasion of the first delivery, that is at the time of the shipping of a product sample.

sag: → *curtaining*

sagging disposition: Unwanted property of a coating material, which becomes visible during its application and drying, through sag formation or → *curtaining*. Sagging can be caused by a deficient formulation, for example too low viscosity, too slow solvent evaporation, or by processing defects like too low temperature and/or too high film thickness.

sagging resistance: Capability of coating materials to form, after application of a defined wet film thickness on vertical surfaces, a well levelling film without producing sagging effects like → *curtaining*. Good sagging resistance normally requires the employment of → *rheological additives*.

salt efflorescence: → *fluosilicate treatment*

salt spray test: Corrosion protection test, according to DIN 50 021 or ISO 7253, which simulates the effects of salt in maritime climates or of road salt in the winter. In a salt spray chamber the test panels are exposed to an artificially produced salt spray fog at elevated temperature, as a rule for a period of 3 to 10 days (yet by agreement also longer periods are possible). The coated test panels are usually scored first, in order to evaluate better rust sub-surface migration on damaged spots. Besides sub-surface migration, also rust formation on the total coated surface and blistering according to the number and size of the blisters, are evaluated.

sand blasting: Collective term for different possibilities of removing old coating layers or rust and scale by means of hard grains (abrasives), which are "shot" at high speed onto the surface to be cleaned. The employed types of grains can be subdivided into "reusable abrasives", which are collected, cleaned and lead back to the process (for example, steel grit, steel shot, corundum, glass beads), and "disposable abrasives" (slag), which are eliminated after use. The real "sand blasting" with silica sand is now prohibited, due to the risk of silicosis. In terms of working methods, one differentiates between "dry blasting", where the abrasive is blasted using compressed air or hurled by a bladed wheel; "damp blasting", which works with moistened abrasive to reduce dust formation; and "wet blasting", where water at a pressure of about 2000 bar is used to transport the abrasive. The different variants of sand blasting work very efficiently and are employed above all in building reuse, heavy duty corrosion protection and infrastructure protection.

sanding: Removal of surplus solid material by rubbing with grained or fibrous abrasives (→ *sanding paper*). In the coating technology, sanding is important in the first place as pre-treatment method for many coating substrates, either in order to level their surface, as in the case of wood, or to improve the coating adhesion by increasing the interfacial contact, as on metal and plastic substrates or on old coating layers. In a multi-coat system, sanding fulfils the same duties, due to the requirement that the already applied layers, for example of putty or surfacer, have to be worked in a way that the next layer to be applied show an optimal adhesion and a satisfying surface.

sanding dust: Dust, generated during sanding processes. Respirable sanding dust can cause diseases and therefore it has to be either extracted by exhaust units or adsorbed by means of dust masks. Before beginning coating application, sanding dust must be completely eliminated, because it would provoke speck formation and gloss reduction in the coating. Besides, during staining of wood, sanding dust situated in the wood pores could lead to "greying" of the colour. The elimination of sanding dust can be carried out by brushing, blowing or sucking off. In addition, in the automotive industry the employment of anti-static clothes is customary.

sanding machines, manual: Vibrating grinders to sand large surfaces. They are used above all in the automotive refinish and in the commercial vehicle sectors to remove old coatings or to sand surfaces which have been coated with putties or surfacers. They can be equipped with sanding paper for dry as well as wet sanding.

sanding paper: It is the most frequently used abrasive. According to the field of employment, it consists of a support of paper in the form of strip, disc or sheet, on which very hard grains of corundum or silicone carbide are fixed by means of a resin. If the paper is additionally impregnated with a water resistant resin, it can also be suitable for → *wet sanding*. The figure printed on the reverse of the sanding paper indicates the grit, expressed as number of grains per square centimetre . The littler the figure, the coarser the sanding paper. Important is also the so called "dispersion", which indirectly expresses how large the interspaces between the grains are. If the paper surface is covered less than 60 % with grains, one has a so called "open dispersion". In the case of more than 60 % it is a "closed dispersion". An open dispersion is advantageous if the materials to be sanded are soft like wood or show at the beginning a sticky surface like polyester putties (→ *inhibitory effect of oxygen*). In these cases, the open dispersion lowers the risk, that the interspaces of the sanding paper are filled with sanding dust or sticky components.

sanding properties: → *grindability*

sand mill: Antiquate variant of the attrition mills (→ *bead mills*). It was an open system, which used as grinding media Ottawa sand, that is a kind of sand with particularly uniform spherical grains. Due to the almost complete depletion of the sand sources and the development of considerably more ecofriendly closed systems, sand mills practically are not employed any more.

saponifiability: Instability of esters against the effects of alkaline agents. This term has its origin in the soap production, starting from natural fats and oils, in which the ester bonds of the used raw materials are decomposed by boiling with caustic soda or potash lye. As reaction products, sodium or potassium salts of the → *fatty acids*, which are called soaps, and glycerine are generated. Referring to this process, in chemistry the decomposition of all kinds of esters into metal salts of the acids and alcohols, by effect of alkaline substances, is generally called "saponification". The term saponifiability is often used synonymously for insufficient chemical resistance. In the coating technology it refers above all to binders and plasticisers containing ester groups which are not structurally protected (steric hindrance).

saponification value: Chemical value, which indicates indirectly the number of ester groups in a film former. The evaluation is carried out by → *titration* with hydrochloric acid, after having let boil the substance in a surplus of potassium hydroxide solution under reflux, in order to saponify it completely.

saturated atmosphere test: It ranks among the most frequently employed corrosion tests. According to DIN 50 021 or ISO 6270, the test pieces are hanged in a closed climate chamber at elevated temperature (mostly

40 °C) above an uniform water surface, so that steadily water vapour condenses on them. Different test variants can be chosen: constant atmosphere or alternating heat/humidity atmosphere, with or without regeneration phases for the specimens.

scale: Black bluish, hard layer consisting principally of a brittle mixture of iron oxides, which is generated during hot rolling or glowing of steel. Before the coating of the steel work pieces, the scale layer has to be completely removed. Otherwise it would affect very negatively the coating adhesion (→ *pretreatment of substrates*).

scattering power: → *light scattering*

scratch resistance: Important property of coats, which, during their use, are frequently confronted with mechanical exposures. This happens, for example, in the case of automotive or domestic appliance coatings, which are regularly cleaned by scouring while they are not permitted to loose their gloss. The scratch resistance of coatings depends in the first place on the structure and cross-linking density of the employed film formers (→ *hardness*); but smoothness and elasticity play an important role as well, concerning the mechanical vulnerability of the coating surface (→ *slipperiness*).

scratch test: → *hardness measurement*

screen analysis: → *sieve analysis*

screening: → *filtration*

scrub resistance: → *abrasion resistance*

sealer: It is applied as sealing interface layer to prevent components of the substrate migrating into the topcoat. This may be the case with substrates of polyvinyl chloride or other plastics containing plasticisers, since there exists the danger of plasticiser migration. But also during the application of alkyd-resin-based coatings on certain type of wood, some natural resin components of the wood may cause a drying retardation or even prevent drying, if the wood substrate has not been sufficiently sealed before.

sediment: Solid deposit on the bottom of a container, caused by the separation of pigments and fillers (extenders), during the storage of a coating material. The reasons for sedimentation is mostly found in the insufficient suspension behaviour of the solid particles in the coating, but at times in defective wetting of the pigments as well. The only remedy is provided by the addition of suitable additives (→ *anti-settling agents* and *wetting agents*).

settling: Tendency of certain pigmented coating systems to form, during storage, a more or less compact sediment. The sediment consists of pigment or filler (extender) particles, enwrapped by film former which, due to their density, in the course of time sink to the ground. If the film former wrapping around the particles is stable and the mechanism of electrostatic rejection is sufficiently pronounced (→ *zeta potential*), the sediment can normally be easily stirred up and thus it represents no real disadvantage. Conversely, the so called "solid sediment" may have a negative effect on the application properties, because it includes the risk that, in spite of intensive stirring up, the pigment particles cannot be dispersed any more without speck formation. The cause of the solid sediment is mostly instable wetting of the pigment particles, often in combination with insufficient addition of → *anti-settling agents*.

shear rate: → *viscosity*

shellac: → *natural resins*

shop primers: They can have different compositions, depending on the type of substrate upon which they are applied. Shop primers serve to insure temporary corrosion protection (as a rule up to 6 months), until the time the workpieces can be processed. Before the final coating of the workpieces, shop primers are removed in the majority of cases (for example in the case of spare parts for car repairs).

short oil: → *alkyd resins, short oil*

shrinkage: Process, which can lead to coating damages as well as to defined required properties, depending on whether it takes place voluntarily or not. Among the coating damages rank shrinkage phenomena due to degradation of the film by weather exposure, or crevice formation in too thick radiation curing layers. Conversely, a "volitional" film shrinkage is planned in the formulations of metallic base coats for two-coat systems. Due to the high solvent amount (or water in the case of water borne coatings), the film of those systems shrinks considerably during the drying and therefore the aluminium or pearlescent pigments are "pressed" into a horizontal position (→ *two-coat metallics*). This is necessary to achieve an optimal effect.

sieve: Netting of fibres or metal with a defined mesh size (→ *filtration*).

sieve analysis: Method for separation of mixtures of granular substances, for example pigments, fillers (extenders) or powder coatings, into different particle size fractions. The fractionating takes place by means of sieve piles, which consist of several sieves accumulated on each other. The sieves have to be arranged top down with decreasing mesh size and they are mostly connected with an eccentric, which put the sieve pile into a vibrating movement. In order to facilitate the sieving process, the sieves are often equipped with spheres, which are put into vibrations and so avoid plugging up on the sieve surface. The material to be fractionated is put onto the upper sieve and passes the different sieve surfaces top down; in doing so, the grains, which are larger than the mesh size of the respective sieve, are held back from this. Thus on the sieve surfaces, fractions with different particle sizes are obtained, which can be weighed to determine the particle size distribution of the mixture. According to the composition and fineness of the mixture, the sieve analysis can be carried out either by a dry or a wet method. In the case of a wet analysis, the particle size fractions have to be dried before weighing.

silica flour: Filler (extender) with high hardness (7 according to Mohs), therefore suitable for applications where the abrasion resistance is particularly important. The main application fields are road line paints and floor varnishes.

silicic acid esters: They are organically modified derivatives of the pure ortho-silicic acid, which cure according to a cross-linking mechanism similar to that of the water glasses (→ *water glass paints*). By partial esterification of the OH-groups of the silicic acid with alcohols, as a rule ethanol, the polarity of the silicic acid is so far much reduced, that sufficient solubility in low molecular alcohols, esters and ketones is provided. Thus, the drying velocity is considerably increased. Addition of water to the solvent mixture makes possible a hydrolytic decomposition of the ester groups, after application, and the separation of the alcohol. The restored OH-groups of the silicic acid react with one another, producing water separation, until a silicone dioxide film is formed (→ *silicification*). The main field of employment of the silicic acid esters is the heavy duty corrosion protection, above all the formulation of → *zinc dust primers.*

silicic acid, precipitated: → *flatting agents.*

silicic acid, pyrogenic: → *high dispersed silicic acid*

silicification: Cross-linking process, which takes place by condensation in the almost completely inorganic so called silicate binders, like potassium and sodium water glasses or ethyl silicate. Thereby, the primarily formed ortho silicic acid becomes in the course of time increasingly water poorer (aging of the silicic acid). As a consequence of the water separation, the silicone atoms link more and more with each other by means of oxygen bridges, until at the end of the process only three-dimensionally cross-linked silicone dioxide remains behind. This is characterised by high corrosion resistance and outstanding mechanical properties. Therefore, the above film formers are predominantly employed in the heavy duty corrosion protection.

silicone dioxide: Colloquially also called "silicic acid", since it can be generated from the ortho-silicic acid by water separation (silicification). In nature, it is predominantly found as quartz and in this form it is employed as filler (extender) in the coating industry (→ *silica flour*). However, synthetically produced silicone dioxides, that is pyrogenic and precipitated silicic acids (→ *silicic acids, pyrogenic* and *flatting agents*) are much more important for the coating technology.

silicone oils: They are mostly short-chain → *polysiloxanes* with linear structure. In the coating technology they play an outstanding role as surface active additives. The low molecular types are above all employed as levelling or anti flowing agents as well as slip and anti-graffiti additives. Their effect is based, on the one hand, upon the incompatibility with the coating systems; on the other hand upon their low surface tension. Thus, in the wet film they are repelled to the surface, where they largely distend (spreading). Doing so, they form a uniform stretched "skin", which levels the surface. Conversely, the higher molecular types meet on the surface in the form of associates, which provoke many little craters. This effect is used for the production of → *hammer finish paints.*

silicone polyester resins: Polyesters, which are modified by a chemical reaction with silicone resins. The addition of silicone resin gives the polyesters a significant improvement in terms of weather resistance. However, this causes also a significant increase in price; consequently, these products, as a rule, are used for special applications. Their main field of employment is the coil coating of façade elements for skyscrapers, for which very long guaranty times are required.

silicone resins: Three-dimensionally curable, organically modified → *polysiloxanes.* They contain reactive OH-groups. The reaction of part of these groups with each other leads to the formation of larger molecules with resin character, which, after coating application, can be later definitively crosslinked by means of the remaining reactive groups and by stoving. The presence of organic groups in the molecules confer on these rather inorganically structured resins a certain compatibility with other organic raw materials as well as good solubility in organic solvents. However, it is worth mentioning that silicone resins, similar to other polysiloxanes, show a tendency to incompatibility and therefore they have to be considered critical problems in coating formulations. Complete compatibility is only achieved by chemical reaction with other film formers (→ *silicone polyester resins*). Due to their water repellence quality, silicone resins have outstanding weather resistance. Additionally, they show the highest heat stability of all coating film formers. Coatings on the basis of silicone resins are durably heat resistant up to 300 °C and temporarily they endure peak values of more than 500 °C. Since their price is relatively high, silicone resins are only used for special applications, where the described properties are essential. Examples of such fields of employment are: architectural protection coatings, electrical insulation segment and the coating of cookers, ovens, exhaust pipes and similar products.

silicones: → *polysiloxanes*

skinning: Special problem of oxidative drying systems and water-borne coatings. In the case of oxidative drying coatings (→ *alkyd resins*), a cross-linking, by means of the oxygen present in the container, may take place on the coating surface already during storage and therefore may cause skin formation. This is more probable in containers which have already been used and are not completely full any more. In order to avoid that, so called → *anti-skinning agents* are used, which temporarily impede the cross-linking reaction. In the case of water-borne coatings, skinning can be caused by drying of the coating surface. The so generated skin is not soluble any more and can provoke speck formation as the outcome of stirring processes. Thus, in the handling of water borne systems one has to pay ultimate attention to carefully close the containers employed.

skull and crossbones: From GHS prescribed danger symbol for very toxic and toxic hazardous substances.

slip additives: They reduce the frictional resistance of a coated surface by improving the surface smoothness (→ *silicone oils*). Doing so, also blocking and scratch resistance are normally increased.

slipperiness: A large number of important processing and application properties like anti-blocking effect, stackability, scratch resistance and anti-graffiti properties depends on it. Slipperiness is reached by use of special additives, which reduce frictional resistance by increasing the smoothness of the surface. For this purpose, additives based on silicone but also different wax types are employed.

slurry process: It is used in the production of heavily-filled water borne dispersion paints, like interior and exterior wall paints. Differently from the dispersion processes in conventional systems, where the pigments and the fillers (extenders) are incorporated into a binder solution (→ *mill base*), in the slurry process only water with some additives is used as dispersion medium. The addition of the film former, in this case of the water borne dispersion, takes place by slow dissolver speed, only after the intensive dispersion of all pigments and fillers. The pigment suspension produced by this variant of dispersion is called “slurry” and it gives its name to the whole process. There are two important reasons for this kind of work mode. First: If large amounts of pigments and fillers are incorporated directly into the dispersion, one would run the risk that the dry powder would absorb very quickly the water of the dispersion, which may possibly break (breaking of a dispersion means the coagulation of the resin in the aqueous phase). The second and perhaps more important reason is the high susceptibility of dispersions to shear forces, which can also lead to breaking. Therefore, the dispersion is added to the slurry only after termination of the dispersion process and on the basis of careful stirring conditions.

smear point: → *yield point*

soft resins: Low to medium molecular, physically drying resins with remarkable plasticising properties. The low molecular, linear structured types are also called “polymer plasticisers”. As a rule, they do not show a real → *gelling power* and they are only migration resistant due to their molecule size.

solids content (non-volatile matter): It expresses the residue of a weighed amount of coating material, which remains after evaporation of the solvents, and possibly of separated reaction products, after drying or stoving at the conditions required by the processing specifications. The solids content is an established parameter for the quality evaluation of coating materials. Among other things it serves to calculate the spreading rate and the VOC (→ *VOC-directive*).

solubility: Capability of a substance to let turn itself into a solution by means of solvents. In the coating technology this normally concerns only film formers and some additives. Solubility depends on the type of the employed solvent (→ *dissolving power*).

solution, colloidal: Here the particle size of the dissolved substance is located between 0.01 and 0.1 µm (→ *solution, true*). The state of association of the molecules or ions is higher than in a true solution. Normally, colloidal solutions contain smaller coils or crystallites of the dissolved substance, which do not cause a marked light scattering but are able to produce, in the almost clear solution, a visible bluish opalescence.

solution, true: This term refers to a state of solution, where the particle size of the dissolved substance is below 0.01 µm. Since the dissolved substance appears in a true solution practically in its finest state of dispersion, that is either molecular or dispersed in the form of ions or finest aggregates, it is not able to refract the light. Therefore, true solutions are basically clear (→ *emulsion*).

solvation: Absorption of solvent molecules on the surface of dissolved particles (atoms, ions, molecules). The so formed complexes are called "solvates". In the case of water, employed as a solvent, the term hydration is commonly used. The solvation of polymers only succeeds if they are not cross-linked and not crystalline. During the solvation of polymers, the solvent penetrate into the "coils" of linear molecules and let them swell. At the same time, the solvent molecules increasingly solvate (encapsulate) the polymer molecules and neutralise their intermolecular forces of attraction, provided that the employed solvent is a "real solvent" for the mentioned system (→ *dissolving power*). When the formation of the so called "solvation wrapping" by means of the solvent absorbed around the polymer molecules is concluded, the molecules are no longer subjected to mutual attraction and are therefore free to move. Thus, the solid film former is turned into a liquid film former solution, which can be used for manufacturing as well as for processing the coating materials.

solvation wrapping: → *solvation*

solvent-free: Strictly speaking, at the moment only the powder coatings and a few water-borne coatings rank among the really solvent-free systems. Here and there also radiation curing systems and some two-pack systems are described as solvent-free, because they do not contain solvents in the true sense of the word but only so called → *reactive thinners*. The latter assume the function of the solvents, concerning the viscosity regulation, but different from the solvents they are chemically linked in the film matrix, so that they are not able to evaporate any more after the coating curing. However, during application they also cause a certain emission, whose intensity depends on the curing time of the coating material. Referring to UV-radiation and electron beam curing systems, the curing times are so short (a few seconds to a split second), that the term "solvent-free" may be acceptable.

solvent, latent: Often employed as thinner component or → *extender*. It is a solvent, which is only able to solve a defined film former in a mixture with other solvents. This is explained because the addition of a second solvent, which could also be a non-solvent, helps shifting the originally non-ideal solubility parameters of the latent solvent in an advantageous direction (→ *dissolving power*).

solvent, real: Solvent, which is able to dissolve a certain film former without the addition of other solvents. A precondition for this is that the so called solubility parameters of the solvent, which represent a measure of the forces of attraction between the solvent molecules, be close to those of the film former (→ *dissolving power*).

solvent resistance: Important property for cured coating systems as well as for organic pigments. Insufficient solvent resistance becomes visible in coating layers by swelling (→ *lifting*). While inorganic pigments like titanium dioxide, iron oxide, carbon black, are absolutely solvent resistant, some organic pigments, in the first place azo-pigments, are subject to selective solubility in certain solvents (→ *bleeding*).

solvent retention: Disposition of dissolved polymers to release some solvent amounts only very slowly. The retention depends decisively on the coordination of the polarity of the solvents and of the film formers. As a rule, the more similar and more polar in their chemical structure the solvents and the film formers are, the more slowly is the evaporation (→ *dissolving power*). Besides, the volatility of the solvents and structural details of the film former molecules also play an eminent role. A high solvent retention has the primary consequence that the mechanical properties of the film can be reached only with a certain delay, since the retained solvent acts like a plasticiser.

solvents: Important but problematic components of many coating materials. In colloquial language use this term refers only to "organic solvents". The main duty of solvents is to transform solid film formers into a liquid state. That is very important for the coating manufacturing (→ *wetting* and *dispersion process*) as well as for the coating application, according to the different application methods. Other functions of the solvents are their helping in the wetting of pigments and coating substrates and the regulation of important properties, during the film formation, like compatibility of the coating components, levelling and gloss. If solvents are only added to the ready coating to regulate its working viscosity, they are normally called "thinners". Solvents are problematic because of many reasons. Since they have to leave completely the film, after the coating application, they are the only coating components which are economically a loss. Besides, solvent emission causes great or moderate damage to human beings and the environment. Effects such as ozone hole, ozone smog and physiological consequences after inhalation of too high concentrations (→ *maximum working condition value*) are well known. Finally, solvents are the main cause of fire and explosion danger in the coating industry and therefore they generate high costs in prevention measures. For several years coating research and development have had as a main target the elimination or reduction of the solvent amount in coatings (→ *low-solvent* and *solvent-free*).

solvent shock: It is the often underestimated result of a mistake during the → *let down* of a mill base. A solvent shock can happen if solvent is added to a highly pigmented and, as a rule, solvent poor mill base all of sudden and without sufficient stirring up. Thus, at the interface between the solvent and the mill base a high concentration gradient is formed, which the solvent tries to eliminate by quick migration into the low-solvent mill base phase. In the boundary layer of the mill base, a too high solvent concentration temporarily appears, with the consequence that the film former wrappings around the pigment particles may be dissolved. Therefore the "naked" pigment particles may attract each other again and form flocculates (→ *flocculation*).

soya bean oil: Semi-drying oil, which is mostly used for the production of white decorative paints. Since it contains preponderantly fatty acids with two double bonds, its tendency to yellowing is relatively low (→ *linseed oil*).

special waste incineration: Waste management process, which is prescribed for extractable residues not suitable for landfill sites. Among them rank all types of liquid coating residues, since they would cause damages to the ground water. Special waste incineration is expensive, because at too low incineration temperatures one runs the risk of formation of highly toxic dioxins. On the other hand, modern plants are always connected with heat exchangers, which permit the recovery of part of the produced energy for heating purposes. An additional problem is represented by the disposal of the produced ashes, since they are often polluted by heavy metals and therefore can only be stored in special landfill sites.

specifications: Inspection and delivery conditions, stipulated between customers and suppliers, which aim to achieve the goal of a satisfying and constant product quality.

specific surface area: → *surface area, specific*

speck formation: Coating damage, which can have multifaceted causes. In most cases it can be linked to insufficient filtration or use of contaminated tools or devices, during the coating application. But sometime speck formation can also be generated by inadequate stability of the coating material, during storage. In this regard examples are incompatibility effects and/or pigment flocculation as a consequence of too poor wetting. In the case of water borne systems, the danger of formation of dried and insoluble crusts should be added. Concerning powder coatings, the agglomeration of the powder particles due to inadequate storage conditions is often the reason for speck formation.

spectrophotometer: → *colour measurement instruments*

spectrum colours: Colours, which are generated by the dispersion of white light into its single wave length bands using a prism or an optical lattice. Each spectrum colour corresponds to a defined wave length and has a maximum chromaticity (saturation or colour intensity). The spectrum colour line represents the outline of the → *chromaticity diagram.*

spinel pigments: Group of synthetically produced inorganic pigments, which in their structure are derived from the mineral "spinel". This occurs in nature as a mixed-phase oxide of the bivalent magnesium and the trivalent aluminium. If these metals in the spinel lattice are substituted by others, for example iron, cobalt, antimony, one obtains coloured variants, which can be used as pigments. Spinel pigments are characterised above all by very high temperature and chemical resistance. Therefore, they are widely used in the ceramic and chinaware industry. On the other hand, since they are relatively expensive, dispersion hard and, depending on the colour, rather poor in tinting strength, they are employed in the coating technology only if the aforementioned positive properties are of the essence. This is above all the case in coil coating systems and highly heat-resistant coatings based upon silicone resins. The most important spinel pigments are: spinel black, zinc iron brown, cobalt blue and cobalt green. A special application of cobalt pigments is the reproduction of the reflection curve of the chlorophyll in camouflage paints. Very similar to the spinel pigments are → *chromium* and *nickel titanium yellow,* which together with these form the group of the "mixed-phase oxide pigments". Chromium and nickel titanium yellow are mainly important for the formulation of lead-free yellow colours.

spot putty: Fine-grained and very quickly drying putty to repair already applied but defective putty layers. Spot putties are mostly based upon nitrocellulose and can be sanded shortly after application, because of their very quick solvent evaporation.

spot repair: Spraying technique in the automotive refinish application. It serves to eliminate locally limited coating damages, avoiding the recoating of a complete body part. The coating material is sprayed, in an increasingly diluted form, from the middle of the damaged area outwards into the faultless surrounding regions, so that, due to the wedge shaped transitional zone, no visible borderline is formed.

spray application: It includes different processes, in which by means of compressed air, of hydraulic pressure on the coating material and/or of an electric field, atomisation of the coating into many little droplets takes place. The droplets fly, under the influence of their own kinetic energy or along the field lines to the object, deposit there and form an uniform film. The particular advantage of the spray application is the high quality of the coated surface. One differentiates between the "pure", the "electrostatic assisted" and the "electrostatic" spray application. The pure spray application includes the →*high pressure spraying*, the low pressure spraying (→*HVLP*), the →*airless* and the →*airmix spraying*. In this regard, one works with compressed air or direct pressure on the coating material. The electrostatic assisted processes use in principle the same techniques, but in addition they employ electrodes, which electrically charge the droplets leaving the spray nozzle. By the guidance effect of the electric field the coating droplets can reach the object in a more targeted manner (→*overspray*). Referring to the electrostatic application processes, the atomisation takes place without pressure exertion, exclusively by the centrifugal effect of rapid-rotation atomisers and under the influence of the electric field (→*electrostatic application*). The main fields of spray application are: automotive refinish and commercial vehicles coating, automotive coating OEM and also some industrial coating processes (for example domestic appliance coating).

spray application processes: →*spray application*

spray booth: Closed room, where the coating of workpieces takes place by means of spray application. Spray booths have to be specially equipped, to avoid negative effects on the coated surface by excessive spray dust formation (→*overspray*) or normal dust, and to eliminate in a safe way the generated dangerous aerosols and solvent vapours. Modern spray booths dispose of efficient ventilation and extraction systems, whereby the air stream is always lead downwards. Thus, the overspray and the solvent vapours, which concentrate at ground level due to their high density, are not whirled up. The bottom of the spray booth consists partly of a grate, lying above a wet rinsing surface or filter mats. The overspray particles, which are dragged downwards by the extraction system, pass the grate and are wetted by the circulating water and carried away. The solvent vapours migrate with the exhaust air either to the catalytic or to the thermal incineration, or to the absorption device. In the case of older spray booths, the vapours may also be directly conducted to the chimney. The cleaned air is enriched with new fresh air, eventually heated by means of a heat exchanger and is lead through filter mats, in order to eliminate dust, back to the spray booth. During the application of waterborne coatings and depending on the quality requirements, the spray booth has to be air-conditioned, that is the values of the temperature as well as of the relative humidity are only permitted to fluctuate within a certain range (→ *relative humidity*).

spray dust: → *aerosol*

spray guns: Devices for the application of coating materials. Depending on the application methods, different variants of spray guns are available. A detailed description of these devices is found in and explained by the following terms: → *airbrush technique, airless spraying, airmix spraying, high pressure spraying* and *electrostatic application*.

spraying automatic devices: Application devices, which do not depend on human manual guide. Their employment permits on the one hand high reproducibility of the application results and on the other hand major improvements in the workplace conditions of the spraying application sector with high physiological exposure. Spraying automatic devices have now become indispensable in many modern coating processes. Their range reaches from simple, non-programmable automatic coating devices, whose spray element executes always the same lifting and oscillating movements, to user-programmable coating robots with manoeuvrable arms, able to imitate practically all movements of a human arm. Between the two there are several variants of partly programmable, static devices whose spray elements can cover different distances along an axis and therefore be suitable for the "wedge application" on test panels (fingerprint coating), as well as automatic devices and robots, performing oscillating movements and able to "travel" with the workpieces along a travel axis in the automotive coating lines.

spraying pattern: This term describes the shape and size of the coated surface, to be obtained when the spray gun is pointed at a spot of a test panel or cardboard and briefly activated. According to the actual adjustment of the spray gun, the spraying pattern can vary between a round shape and a more or less oblong oval. The adjustment of the wished spraying pattern takes place by change of the compressed air supply through the horn passages, which is situated laterally of the nozzle, as well as of the total air pressure. The horn passage air is able to shape the round-section jet into a more or less elongated oval. The longer the oval is, the better are the coating atomisation and uniformity of the coating layer. At the same time, the spray jet becomes drier and that may lead to poorer levelling. This can possibly be remedied by increasing the material supply in the spray gun. In the case of metallic coatings, the optimal adjustment of the spraying pattern depends to a high degree on the required effect. A good painter has the capability to correct, to a certain extent, possible colour mistakes by adequately changing the spraying pattern.

spraying station: Box-shaped, smaller semi-spray booth, which is open to one side and is mainly suitable for laboratories and technological application departments, but also for some industrial uses. Differently from the big and closed devices (→ *spray booth*), in the spraying station the routing of air flow takes place from the front open side to the interior rear-area of the box, that is strictly away from the worker. Before leaving the box, here too the exhaust is cleaned by filter mats and, if necessary, lead to incineration. The spray rooms, where the spraying stations for water borne systems are, have to be air conditioned like the corresponding spray booths in case of highperformance applications. The spray application can take place manually as well as by means of → *spraying automatic devices*. In spraying stations for powder coatings, the exhaust air can be lead back to the box, since it is completely solvent-free.

spraying thinner: Mixture of solvents, which serves to regulate the viscosity at the required value and to achieve the necessary drying properties, during the spray application of coating materials.

spreading: Tendency of liquids to distend themselves willingly on a solid surface. The spreading is the precondition for the wetting of the substrate by the liquid and depends on the respective polarity conditions (→ *wetting* and *surface tension*).

spreading rate: Also called yield. It expresses how many square meters of a surface can be coated with a kilogram of a coating material in the adequate thickness to fulfil the required film properties. The spreading rate depends on the one hand on the coating composition (solids content, density of the coating components); on the other hand it is also influenced by the application technique.

stabilisation (of dispersions): Implementation of measures, which prevent the reversion of a successfully concluded dispersion process. In the coating technology, this concerns the → *dispersion process* of pigments as well as the manufacturing of the water-borne dispersions. Relating to the pigments, it is important to counteract a subsequent → *flocculation* of the finely dispersed primary particles and aggregates. To do that, the mechanisms of the "electrostatic repulsion" and/or the "steric stabilisation" are commonly used. In the case of the electrostatic stabilisation, the pigment particles are kept at a distance from each other by means of self-repulsion, caused by like electrical charges which either have already arisen on the pigment surface under influence of the surrounding medium or are produced by the addition of wetting agents. The steric stabilisation is based on the effect of molecular segments of the wetting agents, which extend from the pigment surface into the binder like stretched arms and so keep the other pigment particles at a distance. Concerning water-borne dispersions, a "breaking" (coagulation) can be avoided by the addition of emulsifiers and stabilising protective colloids but also by Coulomb-repulsion.

staining: Colouring of wood by means of special colorant solutions, so called wood stains, in order to achieve refined effects. There are water-borne as well as solvent-borne wood stains. The application can take place by means of brush, sponge or dipping. The stained wood can subsequently be protected by a clear coat.

stainless steel: High-alloyed steel with more than 12 % chromium ratio. Above all stainless chromium-nickel-steels are widely found in many technical sectors and in the furniture design. As substrates for the coating application they only play an insignificant role, because very seldom they are to be coated.

standardisation: Process to achieve the strict compliance with specified quality criteria. The first step to standardisation is the determination of obliging samples of the materials to be tested. This happens often by agreement with the customers. However, in some cases, for example regarding mixing systems for the automotive refinishing, it can also be an unilateral, internal determination of the coating manufacturer. During the testing of new material batches, a standardised test method generally requires a direct comparison with the corresponding standard specimen. Depending on the quality requirements of the coating material to be tested, this can take place in a very complex way. In the coating technology, measures of standardisation mostly concern the colour matching. Referring to this, the most sophisticated systems are the mixing systems for the automotive refinishing, since the customer paint shops mix their colours by means of a range of readymade mixing coatings and using database formulations without any possibilities to correct them. Thus, the accuracy of colour formulations depends eminently on the exactitude of the used mixing coatings. In this regard, the standardisation process begins already with the careful selection of the pigment batches to be used and ends with the very complex final testing of the produced mixing coating.

standard sample: Reference sample for the evaluation of standardised coating materials (→ *standardisation*).

stand oil: "Thickened oil" (also bodied oil), derived from a drying oil, for example linseed oil, by heating at 260 °C in the absence of air. In this process, a partial cross-linking of the oil molecules with each other takes place by polymerisation of their double bonds, that is without assimilation of oxygen. The production of stand oils on the one hand aims at increasing the viscosity of the oils, which otherwise would be too low to permit a good processing; on the other hand at a better yellowing resistance, due to the poorer oxygen assimilation. A second variant of thickened oil is the "blown oil", where the oxidative cross-linking is partly anticipated at higher temperatures and by the supply of oxygen.

stearic acid: Saturated monocarboxylic acid with eighteen carbon atoms. It is contained in little amounts in many natural oils and is, among other things, employed as grinding agent in the production of aluminium bronzes (→ *leafing bronzes*).

steel: It is still the most important metallic material. It is produced by smelting of predominantly oxidic iron ores in the presence of coal, whereby the oxygen from the iron oxides is bound by the coal in the form of carbon dioxide, while the reduced iron is released as pure metal. In the first process stage, so called "pig iron" with a carbon content of about 15 % is produced, which cannot be technologically used, due to its brittleness. Only a reduction of the carbon content under 4 % leads to valuable materials. This can be achieved by different processes. Between 4 % and 1.5 % carbon content, cast iron materials are obtained. Under 1.5 %, one gets unalloyed steel. In order to improve some properties (for example tensile strength, hardness and corrosion resistance) the ordinary steel, which is often colloquially simply called iron, can be alloyed by the addition of other metals like chromium, nickel, and molybdenum. "Low-alloy steel" contains up to 5 % foreign metal content, "high-alloy steel" beyond it, but often, under the designations high-grade steel and stainless steel, up to 30 %. As substrates to be coated, above all unalloyed and low-alloy steels are interesting, since only these are subjected to corrosion and therefore demand appropriate protection measures.

sticky tape detachment: Non-standardised accentuation of the → *cross hatch adhesion test*, which can be encountered above all in specifications of the automotive industry. Thereby, according to the specification conditions, a strip of adhesive tape is attached to the cutting area of the coating surface and than quickly detached by a sudden traction, vertically to the surface. Thus, delamination processes of the coating layers from the substrate or of the single layers from each other are made better visible.

stirrers: Devices, which are used in the coating technology to homogenise solutions, mix pastes, incorporate pigments and fillers (extenders) into the binders or simply to stir up ready coatings. Stirrers differ from → *dissolvers*, because normally they work at low speed (seldom above 1500rpm/revolutions per minute), their peripheral velocity often is not continuosly variable, and above all because they only dispose of non-toothed disks or propellers (propeller stirrer). In the stage of the coating production, where the pigments and fillers (extenders) are incorporated into the mill base, the dissolver assumes the function of a stirrer, in order to avoid a change of in the dispersing unit. But in such cases the process does not take place under dissolver conditions but with slowly rotating disks.

stone chipping protection: Demand to be met in all systems of the automotive coating and the coating of commercial vehicles. Particularly stone chipping endangered areas like underbody and wheel guards are protected by well adhering, hard-elastic materials, mostly based upon polyvinyl chloride → *plastisols* or polyurethanes. For all remaining areas, the surfacer assumes the stone chipping protection function, due to its reinforcement with special fillers (extenders).

stone chipping test: For the quick evaluation of stone chipping resistance, numerous test methods are available. The best known among the standardised methods are the "pecking chicken method" according to DIN 55 995, in which the coated surface gets a single impact by shooting of a ball (Ladstädter-airgun or BMW-mono-impact), and the shot blast test according to DIN 53 154, in which 1000 steel balls with 10mm diameter drop onto an oblique lying coated panel, falling from a defined height. Besides, there exist different test variants, according to the specifications of several automotive manufacturers.

storage stability: On storage stability depends the period of time, during which the coating material can be stored, without change of its processibility and/or of its technological properties. Beyond this time limit, normally no liability about possible product defects is assumed. Information about storage stability are an inherent part of technical data sheets and delivery specifications. Referring to this, it is very important that the storage conditions be exactly set (for example, temperature and moisture.).

stoving: Technology, suitable to carry out the chemical cross-linking of → *stoving coatings*. These contain film former components, which can chemically react with each other only at higher temperatures. However, the completeness of the cross-linking depends on the → *stoving tempera-*

ture as well as on the stoving time. For the stoving processes, different variants of driers can be employed. Common are the convection driers, also called circulating air driers, the IR-driers (→ *infrared drying*) or combinations of both. In the convection driers, heated air circulates first to the object and then, after having donated a part of its heat, back to the heat exchanger. Since, doing that, it enriches generally on solvent vapours and separation products from the cross-linking reaction, it is steadily mixed with fresh air, whilst a part of the polluted air leaves the drier as exhaust.

stoving coatings: Coating materials, whose curing is caused by a chemical reaction, which takes place at high temperatures (→ *stoving temperature*). Stoving coatings are typical one component systems, which can be stored without any problem at room temperature and only after application undergo a chemical reaction in suitable stove furnaces. In most cases, the cross-linking takes place by a condensation reaction with separation of low molecular reaction products like water, alcohols or formaldeyde.

stoving temperature: It is the temperature which has to be reached, depending on a defined stoving time, to achieve a complete and failure-free curing of the coating material. The usual stoving temperatures vary between 110 and 220 °C.

stress: → *viscosity*

structured paints: Coatings, which, due to their rheological properties and/or the effects of special additives, after their drying show a structured decorative surface. Among the oldest structure paints rank the → *wrinkle finishes*. Modern structured paints contain relatively large amounts of rheological additives, which produce in them a marked → *pseudoplasticity*, so that the coating material during application is hardly able to level. The so formed surface structure can be additionally influenced by increasing the solids content and the use of hot spraying techniques. A special and very popular variant of structured paints is the → *hammer finish paint*.

styrenated alkyd resins: They are long oil alkyd resins, which are modified with styrene (→ *alkyd resins, long oil*). What takes place is a linking of polystyrene-chains at double bonds of the drying oil. Thus, both the physical drying and the resistance against hydrolysis and saponification of the alkyd resins are improved. On the other hand, the solvent and scratch resistance are influenced negatively. However, that does not represent a real disadvantage in the use of primers, because they are always mechanically and chemically protected by a top coat. Therefore, oxidative drying corrosion protection primers represent the main field of employment for the styrenated alkyd resins.

styrene: Low viscosity liquid, inclined to polymerisation, whose correct name is “vinyl benzene”. In the coating technology, styrene plays on the one hand an important role as so called → *reactive thinner* for unsaturated polyesters; on the other hand as monomer for the production of acrylic-styrene-copolymers for exterior wall paints (→ *wall paints, exterior*) and top coats and finally as modification component for alkyd resins (→ *styrenated alkyd resins*).

subsequent cross-linking: Mostly voluntarily carried out and additional cross-linking of an already hardened or physically dried system (→ *drying, physical*), by means of chemical groups which show a long running reaction. Corresponding examples are the oxidative subsequent cross-linking of the physically drying cyclised rubber, due to the double bonds present in its molecules, or of stoving systems containing → *dehydrated castor oil alkyds*. The subsequent cross-linking causes normally an improvement of the chemical resistance and of the mechanical stability in the coating film. However, the elasticity generally worsens.

substrate: Term used in the coating technology to express the ground for coating application. What is meant is not always the bare ground which exists before application of a complete multi-layer coating system (metal, plastic, wood etc.), but some time also an already coated ground, which must get a further coat. For the additional coating layer, the substrate is in this case the dried or cured coating material on the object. Properties like adhesion, gloss and optical appearance of the coating depend preponderantly on the substrate and its pre-treatment.

sub-surface rusting: It is the term to express the rust formation which, during weathering tests, begins on a spot of the coating which has been intentionally damaged by a linear scratch. The rust, arising in the scratch trace, spread more or less under the coating layer. The sub-surface rusting degree is evaluated according to ISO 4628.

subtractive colour mixture: Method of colour formation by means of surface colours, that is of substances, which are able to absorb visible light (→ *additive colour mixture*). Since all pigments used in the coating technology, with exception of the pearlescent pigments and liquid crystals (→ *interference pigments*), are pure → *surface colours*, the subtractive colour mixture represents the main mechanism of colour formation in coatings. Here the pigment particles, which are finely distributed in the coating material, act as light filters, absorbing and turning a part of the illuminating white light into heat (→ *light absorption*). This variant of colour mixture is called “subtractive” why the different pigments, imbedded in the coating, can “subtract” different wave length ranges from the white light. The colour, visible to our eye, is formed from the wave lengths of the white light which has remained after the subtraction and is reflected from a scattering coating component or from substrate (→ *light scattering*).

suggested formulations: Also called “starting point formulations”. They are standard formulations, which are mostly developed by producers of coating raw materials and serve as a basis for testing of their own products. Suggested formulations facilitate the introduction of new raw materials into the product range of coating companies, since they offer already tested combinations which can be adopted with only little expenditure made in adaptation work.

sun flower oil: Non-drying oil, used predominantly for stoving coatings.

surface active: Property of different coating additives, which belong to the group of the surfactants. These substances are able to influence the surface tension of phases, which have an interface in common. According to their kind of effect, these surface active substances can be employed as wetting agents, emulsifiers, levelling agents, defoamers, anti-floating agents and for other applications.

surface area, specific: Value, which specifies the total surface of all particles contained in one gram of pigment. It is expressed in m^2/g and it depends on the density as well as on the particle size of the pigment. Its numerical value increases, if the density and the diameter of the pigment particles decrease. The specific surface area states indirectly how easy or hard is to disperse a pigment, during the coating production. The bigger the specific surface area is, the bigger is the total of the forces of attraction, which hold together the particles in the form of pigment agglomerates and the larger is the pigment surface, which has to be wetted by the binder solution (→ *dispersion process*). The evaluation of the specific surface area takes place according to the BET-method, whereby the amount of nitrogen is determined, which the surface of the required pigment is able to absorb. The numerical value of the specific surface area is situated for inorganic pigments between 2 and $20 m^2/g$, for organic ones between 10 and $80 m^2/g$, for pyrogenic silicic acids between 120 and $380 m^2/g$ and for carbon blacks between 50 and 700 m^2/g.

surface cleaning: It is the first step in the pre-treatment processes of all kinds of surfaces, before the application of coating materials. A careful cleaning is indispensable for an optimal coating adhesion (→ *degreasing*). At the same time it can also have a main importance for avoiding coating damages like cratering and edge effects and for achieving the required optical results, for example by staining of wood. Cleaning methods are different, according to the properties of the surfaces. While in the case of metals and plastic the elimination of traces of grease and silicones has to be the focus, mineral substrates require more the removing of slack dirt, dust and chalking layers, and finally wood necessitates the treatment of resinating spots by means of solvents and the careful elimination of abrasive dust.

surface colour: Colour of an object, which is not able to irradiate its own light and therefore it depends on a foreign illumination to show its own colour trait. Conversely, a "self-luminous object" shows its colour by spreading its own radiation (for example sun, neon light, the light spots of a colour-television set). Coatings are always surface colours. Only if they are irradiated by a light, they can transform this in a way to produce a colour impression (→ *light absorption* and *light scattering*).

surface damages: Effects, caused by coating material defects or processing mistakes, which influence negatively the required appearance of the coating surface. Mostly, surface damages are connected with insufficient levelling or with incorrect working, during the pre-treatment or the application. The best known damages are → *orange peel effect* and *cratering*. In the broadest sense, also → *blistering* and specks, caused by unclean coating material and/or application devices, belong to these.

surface protection: It is one of the most important duties of almost all coating systems, whereas, according to the kind and use of the surface to be protected, this term can refer to different meanings. Thus, the term "surface protection" can express the protection of metals against corrosion as well as the protection of metals, plastics and wood against mechanical exposures, the protection of plastics and wood against degradation caused by UV-radiation and the protection of all kinds of substrates against the different chemical influences.

surfacers: → *primer surfacers*

surface tension: Expression of the effect of the cohesion forces at the surface of a substance (→ *cohesion*). The particles of a substance are held together by different kinds of physical and chemical forces of attraction, whose strength depends on the → *polarity* of the substance. The particles, which are situated inside the substance, are surrounded by identical particles, which apply the same forces one upon the other. However, this does not occur in the case of particles at the surface. Downwards and sidewards they are subjected to the forces of attraction of identical particles, yet outwards they develop interactions with particles, which belong to another substance (for example air) and therefore exert other forces of attraction. The bigger the draw to the inside of the substance is, which the particle layer at the surface underlies due to the attraction of identical particles, the bigger is the surface tension of this substance. In the coating technology, the surface tension of liquid coating components is particularly important, because on it depends the → *wetting* of pigments and fillers (extenders) as well as of the substrates to be coated.

surfactants: → *surface active* and → *wetting agents*

suspension: Heterogeneous system consisting of solid substances finely distributed in a liquid. A pigmented coating, whose pigment particles are dispersed in a dissolved binder, represents as a rule a suspension.

swelling: It means the diffusion of solvents into a solid substance and the solvation (encapsulation) of its molecules without provoking a "real" dissolving process. The result of swelling is, that the solid substance becomes softer and more voluminous. During coating application, a weak swelling of an already existing coating layer by the solvents of the applied wet material is generally advantageous, because the interlayer adhesion may be improved. Conversely, a too strong swelling can lead to a debonding of the coating film from the substrate and to wrinkling (→ *lifting*). In the case of metallic coatings, swelling is often the cause of clouding.

synthesis: Chemical process, whereby substances react with each other, that is they link with each other, to form new substances. The synthesis is the base for the artificial manufacturing of substances by chemical means.

synthetic: Artificially produced. In the coating technology this is now true of almost all employed raw materials. An exception are most fillers (extenders), which are preponderantly manufactured by processing of natural minerals, and different solvents which result as fractions in the oil distillation. Concerning the film formers, natural raw materials are either synthetically changed (for example colophony resins, cellulose nitrate, cyclised rubber) or they serve to modify other synthetically produced film formers (for example fatty acids in the alkyd resin manufacturing). However, the largest amount of film formers is generated synthetically. This is also true of all pigments and most additives.

synthetic resins: Collective term for all synthetically, that is artificially, produced film formers (→ *resins*). This term was created at the beginning of the twentieth century, when the first synthetic resins were developed (phenolic, vinyl and urea resins), to distinguish them from the natural resins (dammar, copal, shellac etc.), which had been employed until then.

T

talc, talcum: Important filler (extender), above all for the formulation of putties, surfacers and corrosion protection primers. Talc is exclusively produced from natural deposits and chemically it is magnesium silicate hydrate. Due to its marked platelet shape, it is particularly suitable to formation of → *barrier effects* against moisture diffusion. Its outstanding softness (hardness according to Mohs = 1) promotes its employment in coating materials, which have to be sanded after application, like, for example, putties and surfacers. Besides, in the hardened film the platelet shape of its particles causes durable reinforcement, which becomes visible by crack bridging and improved elasticity. Through the anchorage of the platelets in the sanding scratches of the substrate, the talc particles provide an improvement of the coating adhesion. On the surface, their protruding edges cause a marked matting effect; therefore, talc is also often used as cheap flatting agent. Conversely, the high binder requirement of this filler is in most cases a disadvantage, because it provokes a large increase in viscosity. Also the relatively large particle sizes (the finest talc types have a maximum particle size about 10 up to 15μm) reduce the possibilities of use of talc in materials, which have to be applied by → *wet-on-wet-application.* In automotive surfacers for OEM as well as in cathodic electrodeposition primers, talc is normally replaced by the finer platelet shaped → *kaolin.*

tall oil: So called, according to the Swedish term "tall", which means "pine". In reality tall oil is no real oil, since it does not consist of an ester from glycerine and fatty acids. It is instead a mixture of semi-drying fatty acids (mostly linoleic acid), which occurs as a secondary product, together with → *colophony* and → *turpentine oil,* by the extraction of cellulose from conifers. By means of suitable rectification processes it can be refined and so influenced in its composition, that its properties can be considered now quite reproducible. Due to its low price, it is above all employed in coloured decorative paints.

tank farm: Part of the raw materials warehouse, which serves for the safe storage of large solvent amounts. The storage can take place either above the surface or be subterranean, due to the risk of fire of the solvents and depending on the building structure of the coating plant. In the case of a storage above the surface, outdoor, the tanks have to be protected against excessive heating by sun radiation. In all cases, also by subterranean storage, special sealed basins of steel concrete with an adequate volume capacity have to be installed under the tanks, in order to intercept solvents in case of spills, without danger for the ground water. During all filling operations, the pipelines and the employed vessels have to be earthed (grounded), in order to prevent fire and explosion risks (→ *electrostatic charge*).

temperature: Base physical quantity, which expresses the internal movement condition of the matter. In the case of scientific calculations, one conforms to the official base unit "Kelvin". The zero point of the Kelvin scale is at 273.15 °C and corresponds to the temperature, at which the molecular movement of the matter has reached the absolute immobility (absolute zero point). In the coating technology, temperature plays an important role. It is an essential parameter for the dissolution and dispersion processes as well as for the extrusion of powder coatings (→ *extruder*) and the drying/ stoving processes of coatings in general.

terephthalic acid: Variant of the phthalic acid, where the carboxyl groups are situated in para-position, that is they are opposite each other in the molecule (→ *phthalic acid*). Through this position of the acid groups, polyesters with marked linear (chain-like) molecules can be obtained. Due to their high flexibility and surface hardness, such film formers are particularly suitable for → *coil coating* applications and the coating of electrical insulating wires.

terpenoid hydrocarbons: Substances of vegetarian provenance, whose components are extracted by distillation of the balsam of different conifers. Some of them are used as solvents. The best known representative in this function is the → *turpentine oil*. But also pine oil and dipentene are important, above all as high boiling solvents. Their dissolving power is situated between that of the → *aliphatic hydrocarbons* and of the aromatic ones. Relating to the chemical structure of its components, also the most important natural resin for the coating industry, the → *colophony*, can be counted among the terpenoid hydrocarbons.

test methods: Methods for the evaluation of quality features of raw materials and coating materials as well as to guarantee the compliance with the laid down specifications before delivery of such materials to the customers.

test specifications: Binding agreement about test methods and their parameters for testing of raw materials and coating materials, aiming at comparability of the results. Test specifications are, as a rule, firm components of the delivery specifications, which have been laid down between the coating manufacturer and the customer. They are normally based upon existing standards (for example, DIN- and ISO-standards, or company standards), but they can also be set, diverging from the usual test practice.

thermoplastic resins: Resins, whose molecules have a linear (chain shaped) structure and are not cross-linked. By adequate temperature increases, their mobility can be so elevated, that they begin to "flow". Similar effects happen, if their molecules are sufficiently solvated by means of suitable solvents (→ *solvation*). However, this succeeds only in the case of low molecular types, that is of little molecules. In practice, the sensitivity of most thermoplastic resins to solvents becomes visible by → *swelling*. Differently from → *thermosets*, thermoplastic resins are constantly fusible and can be dissolved or swelled by suitable solvents. Their fusibility makes possible to process them by means of simple shape-giving methods, like for example the injection moulding process, and permits also the use of efficient recycling technologies. Since most thermoplastic resins are produced by polymerisation processes (→ *polymerisation*), through the choice of suitable monomers it is possible to achieve different technological properties in a very selective way. Versatility, simplicity of their processing, low cost and low weight have enormously increased the importance of the thermoplastic resins in the modern technique. This represents, at the same time, a challenge for the coating industry (→ *coatings for plastics*). Examples of particularly important thermoplastic resins are: polyethylene, polypropylene, polyvinyl chloride, polystyrene, polycarbonate, polymethylmethacrylate/ Plexiglas, acrylonitrile-butadiene-styrene-copolymers, polyamide and thermoplastic acrylates.

thermosets: Closely cross-linked polymers. Thermosets do not soften at elevated temperatures and are not soluble any more. They are mostly generated from thermoplastic molecules by means of a chemical cross-linking reaction (→ *thermoplastic resins*). In the coating technology this process takes place during the coating hardening, for example by stoving, by two-pack crosslinking or by UV-curing.

thermosetting: Variant of the coating cross-linking, which takes place by heat addition in circulating air driers or infrared driers (→ *stoving coatings, infrared drying* and *convection drier*). The reaction between the cross-linking partners is as a rule endothermic, that is it needs heat addition until its conclusion. If the heat supply is interrupted too early, the cross-linking reaction remains incomplete with consequences corresponding to the resistance properties of the coating film. Therefore, it matters to fix the stoving parameters, temperature and stoving time, so that the needed heat exchange may take place completely.

thickener: Rheological additive for applications in water- or solvent-borne systems (→ *rheological additives*).

thickening: Unwanted remarkable viscosity increase, which can arise under certain conditions during the storage of a coating material. Its causes can be multiple. Unsuitable storage conditions (for example too high temperatures in the case of unsaturated polyesters or frost in the case of dispersion paints) may lead to thickening, as well as an unstable composition of the coating (for example presence of basic pigments, like zinc oxide, in acid binders or insufficient amounts of amines in water borne systems).

thinnability: Capability of a dissolved system to tolerate defined amounts of thinner without showing turbidity, separation appearances or other failures.

thinner: Solvent or mixture of solvents, which is added to an already dissolved coating system to adjust its application properties. A thinner consists normally of real solvents, but it can also contain amounts of latent solvents and sometimes even of non-solvents acting as solvent extenders (→ *solvent, real; solvent, latent* and *non-solvent*), provided that the latter evaporate more rapidly than the real solvents (→ *extenders*).

thixotropic agents: → *rheological additives*

thixotropy: Rheological behaviour of non-Newtonian liquids, which by an increasing shear time and a constant shear rate show a viscosity decrease (→ *pseudoplasticity*). In practice, thixtropy becomes visible when, by effect of shear forces during the coating application, for example caused by air streams in a spray gun, the coating material first becomes thinner and then retrieves the originally higher viscosity only after a certain time. This has the advantage that, when the coating droplets deposit on the surface of the workpiece during application, they have enough time to spread before viscosity increases again. Thus, it is possible to build up relatively thick coating layers without sagging and → *curtaining* and to get an acceptable levelling at the same time.

through curing: State of complete solidification of a coating film, also in its deepest layers, caused by a chemical cross-linking reaction. The entirety of the through curing depends on the present reaction conditions. Problems, during the through curing process, can arise for example in the case of incorrect drier addition to oxidative drying alkyd resins or if UV-curing systems contain UV-absorbing pigments.

throwing power: Phenomenon, which appears in electrodeposition and electrostatic application methods, because of the course of the electrical field lines (→ *electrostatic application, electrodeposition coating* and *EPS-coating*). In the electrodeposition processes the throwing power effects, that also hollow spaces, which normally impede the access of electrical field lines because of their "Faraday cage" nature, in the course of time are increasingly reached by the field lines. This is based upon the fact that the coating material, which has already been deposited upon the workpiece surface, acts like an insulating layer. Therefore, the field lines are moved to the still conductive zones of the object, for example voids. Also in these an optimal layer formation takes place. A further advantage of the throwing power is the very good edge covering, which is caused by a concentration of the field lines on narrow edges and pointed parts (→ *edge covering*).Thereby, an optimal corrosion protection performance can be achieved just on these risk zones. In the electrostatic application and EPS-coating, the throwing power permits the simultaneous coating of the front and the reverse of an object, since a part of the field lines runs around the earthed object.

tinting: Matching of the colour of a coating material, according to a reference panel, by means of tinting bases or mixing coatings. In the coating production, this activity is performed by collaborators with special skills, the so called "tinters", who in addition to great experience in handling of tinting components, must also possess efficient colour perception.

tinting base: Paste, consisting of a pigment in relatively high concentration and a binder, which is compatible with the coating system to be tinted, including dispersing agents. In the ideal case, the binder is the same contained in the coating system. The pigment concentration, the tinting power and the colour location of the tinting base should be as constant as possible. Without these preconditions, the tinter would be hardly able to estimate the right paste amounts to be used and besides no colorimetric calculation to support the tinting process could be carried out.

tinting strength: It is a measure of the capability of a pigment in a mixture, to impose its colour onto the other mixture components. The tinting strength of a pigment depends much on its dispersion degree. Therefore, in the coating practice often tinting strength development curves are worked out. Thereby, the pigment is dispersed for differentiated periods of time and then it is mixed with a defined amount of white and evaluated either optically or colorimetrically.

titanium dioxide: It is the most important of all white pigments. It can be produced in the crystal modifications of anatase and rutile. For the coating technology almost only the rutile modification is interesting, since it is less susceptible to chalking (→ *anatase*). Due to its superior refractive index of 2.75 (for example zinc sulphide, the white component of the lithopone, has only 2.34), titanium dioxide has displaced almost completely the other white pigments from the coating sector. Two processes exist for the manufacturing of titanium dioxide: the sulphate and the chloride process. In the older sulphate process ilmenite, a mixed oxide of titanium and iron, is used as raw material. The mineral is decomposed by means of concentrated sulphuric acid, thus titanium is dissolved and so separated from iron. Afterwards titanium oxide hydrate is precipitated by diluting the solution (hydrolysis) and is calcinated in rotary furnaces. A disadvantage in this process is the formation of the so called "thin acid", which in the past was poured into the sea, provoking considerable ecological consequences to the environment. Since discharge prohibition has come into force world wide, methods have been developed in order to concentrate the thin acid and lead it back to the process. In the chloride process, the pigment formation takes place in the vapour phase. That has the advantage of a more pure white colour. By reaction with chlorine, rutile sand is transformed into the volatile titanium tetra chloride, which is purified by distillation. After that, the transformation into rutile takes place by means of oxygen. Regardless of the process which has been used to produce the titanium dioxide, subsequently the pigment has to be stabilised against photochemical activity by post-treatment (→ *photochemical degradation*). There are different variants of post-treatment processes. Most frequently, the stabilisation is generated by layers of aluminium oxide and/or silicone dioxide. Highly weather resistant types are coated with up to 15 % silicone dioxide or zirconium oxide. Thus, no direct contact between the titanium dioxide surface and the binder is possible any more and therefore photochemical degradation is prevented. To improve the dispersibility (dissolver grades), a post-treatment with polyols may also take place.

titration: Analytical process for the quantitative evaluation of substances by means of standard solutions. The standard solutions contain suitable reagents in known concentration, which have to react completely with the substances to be evaluated. The standard solution is added to the specimen in portions by means of a dosage apparatus of glass (burette), which makes possible to measure volumes very exactly. The end of the reaction (equivalence point) is mostly showed by the change of colour of an indicator, added to the test solution, or of the standard solution itself. But there exists also the possibility to show the equivalence point using physical-chemical measuring methods, for example the potentiometry. From the volume of the standard solution, added until the equivalence point is reached, the wanted amount of the test substance can be calculated. In the coating technology, the evaluation of different values of film formers by means of titration has a considerable importance. Thereby, one gets valuable information about the composition and suitability to cross-linking of the film former molecules. Particularly important are the →*acid value*, the hydroxyl value, the iodine value, the epoxy value and epoxy equivalent weight and the isocyanate value.

tolerance: It is the fluctuation range, within which in a measuring process a divergence from a fixed reference value can be accepted. A tolerance can be laid down, due to the natural fluctuation of a measuring method or of a measuring device. However, it can also be fixed in an agreement between suppliers and customers.

toluene: Aromatic solvent with high dissolving power in the medium polarity range. Similar to →*xylene,* toluene dissolves alkyd resins, non-modified polyesters, acrylic resins, rubber derivatives and several copolymers. Its employment is reduced because of its disadvantageous flash point (below 21 °C) and the relatively low →*MWC-value.*

toluolsulphonic acid: →*acid curing*

top coat: Exterior protection skin of each multi-layer coating system, which is responsible for the optical appearance as well as important for the mechanical and chemical resistance of the whole coating system. Depending on the requirements for colour and surface quality, the top coat may consist of one or several layers. In the case of plain colours, that is those equipped with hiding pigments, only one layer is normally sufficient. If the requirements are particularly high (for example in the automotive coating), here two-coat systems are increasingly employed, which means that the hiding coating film is additionally protected by a clear coat film. In the automotive technology, metallic colours consists generally of two coats (→*twocoat systems*). So called “pearls”, containing non-hiding interference pigments, need a third film, as a rule white or black, which is applied as under layer, according to the required effect.

toxicity: Property of some hazardous substances, which according to →*GHS* and depending on their danger level, have to be labelled with the danger symbol “human torso” or “skull and cross bones”. Very toxic substances are not employed in the coating industry, but even just toxic ones are now scarcely used (like for example cresol, zinc and lead chromate).

transparency: It arises, if a substance is not able to scatter or to completely absorb light. Thus, the light can completely or partially pass trough the substance. In the coating technology, clear coats are in the first place transparent. Transparency is also generated in pigmented coatings, if the employed pigments have very small particles (particle size below 0.2 μm). Under this particle size no light scattering is possible any more. Therefore, a ray of light, after having possibly adopted the colour of the pigment embedded in the coating, can pass without impediments through the coating film and reach the substrate (→ *wood glazes*).

treatment of sewage: During the processing of coatings, above all in the stage of industrial spraying application, a lot of polluted sewage water can accrue. This traces back to the fact that in wet rinsing spray booth the water is used to wet and remove the coating overspray. These sewages enriched with coating have to be pre-treated, according to the pertinent regulations, before discharging into the public canalisation, otherwise they would cause damages later in the communal biological wastewater treatment plant. The coating residues are almost completely removed from the sewage by means of special coagulation processes. For the remaining organic substances defined limiting values of COD (chemical oxygen demand) are prescribed. Furthermore, some heavy metal concentrations must not be exceeded and the sewage water has to be almost neutral (pH-value = ~7).

tribo-application: → *EPS-coating*

trilene dipping process: In the past an important conventional dipping process (→ *dipping*), it is nowadays a hardly used process, because of its effect on the environment and the physiological risks it poses (→ *chlorinated hydrocarbons*). On the other hand, the trilene dipping process offers the advantage that the coating can be carried out safely at high temperatures, due to the incombustible nature of the used solvent trichloroethylene. This makes possible very short drying times and high production performances. As a modern alternative, water-borne systems are available, although they require much more energy for the drying process.

trimellitic acid: Organic acid with three carboxyl groups. It shows a similar structure to the → *phthalic acid* but it permits, by means of the third carboxyl group, the formation of so called "acid polyesters", which are important for the production of water-borne film formers as well as of → *powder coatings*.

trimethylol propane: The most important among the trivalent alcohols, which can be used for manufacturing of polyesters (→ *polyester resins*). Due to the higher resistance to saponification of the polyesters, which are produced with it, trimethylol propane is mostly preferred to glycerine. Besides, trimethylol propane is important as modification component for → *colophony* and → *maleic resins* and as reaction partner for the trimerisation of aromatic isocyanates (→ *polyisocyanates*).

triple roll mill: Dispersing equipment for paste-like mill bases. In the past it was very important in the coating production, above all for the dispersion of primers and surfacers but also for plain colour top coats. It consists of three horizontal rolls, which rotate with different velocities and in reverse of each other. Their rotation speeds are in the proportion of 1:3:9. The pasty mill base is fed between the first roll (feed roll) and the second one (centre roll) and is transported by the third roll (take-off roll) to the stripper blade. The first and the third roll can be pressed with different pressures against the central one; thereby the slit (nip) width between the rolls, and thus the shear intensity, can be varied. The importance of the triple roll mill in the modern coating technology is decisively less, due to the high solvent emission. However, this dispersing device plays still a relevant role in the manufacturing of printing inks and artist's colours, because here only solvent poor or free mill pastes are employed.

tristimulus method: → *colour measurement instruments*

tristimulus values: They are the absolute values of the red content X, green content Y and blue content Z of a colour. They are calculated from the reflection curve (→ *colour measurement instruments*) by means of the sensitivity curves of the eye (standard colour matching functions) and the spectral composition of the employed light. For practical use, they are converted into standard colour coordinates x, y and z, dividing each of them by the sum of all three (→ *chromaticity diagram*).

trivalent alcohols: Alcohols, which contains in their molecule three reactive OH-groups. These organic compounds are important raw materials for polyester production, which can be later cross-linked by a chemical reaction (→ *polyester resins*). The most commonly used trivalent alcohols are glycerine and trimethylol propane.

trowelling: Application method for all kinds of putties. For it, special tools (spatulas) are used consisting of a more or less flexible sheet of metal or plastic. The filling compound is first applied on the uneven surface and subsequently smoothed by means of the sheet. After the hardening of the putty, the surface is levelled by sanding.

tube enamels: Highly elastic stoving systems for the coating of tubes of metal alloys. Beyond the very high elasticity, necessary because of the extreme deformations during their use, they have also to show an outstanding adhesion even under disadvantageous use conditions (for example bath atmosphere with high moisture and heat). As film formers, stoving →*epoxy resin esters* in combination with melamine resins are generally employed.

turpentine oil: It is the most important representative of the terpenoid hydrocarbons in their application as solvents. It is mainly extracted as balsam turpentine oil, together with colophony and →*tall oil*, during the production process of cellulose from conifers. It is characterised by an intense and typical smell and it shows an outstanding dissolving power, above all to oil paints and varnishes, as well as to long oil alkyd resins. In the coating technology it is mostly employed in little amounts as high boiling solvent to improve the levelling of coating materials. Besides, it is very popular as classic thinner for artist's colours. During its employment, attention has to be paid to its allergic effects, which may lead to skin reactions.

turpentine oil substituted: Mixture of high boiling aliphatic hydrocarbons, which is used instead of the more expensive turpentine oil, above all for application as cleaning solvent and thinner.

two-coat metallic coatings: They consist of a base coat and a clear coat (topcoat). The base coat has exclusively the duty to provide, by a quick fixing of the aluminium platelets and a large film shrinkage during the drying, an optimal horizontal orientation of the platelets and therefore a satisfying metal effect (mirror effect). The clear coat, mostly a two-pack system or a stoving coating, has to protect against weather and light exposure as well as mechanical and chemical influences. In order to improve the light fastness, the clear coat is equipped with light stabilisers. In addition, only very light resistant pigments are permitted to be used in the base coat, since the damaging UV-radiation entering the film is reflected by the metal platelets and thus the pigment particles are more intensively affected.

two-coat systems: It is the coating variant, which is encountered most frequently in reference to metallic colours (→*two-coat metallic coatings*). However, in the automotive industry plain colours are also increasingly applied by the two-coat process, due to the fact that significant advantages in the gloss and the light and weather resistance can be achieved.

two-component systems: Coating materials which consist of a main component, mostly called "base material", and of a so called "hardener component". The two components are stored in separated containers and are mixed with each other, according to a prescribed ratio, only shortly before their use. At that time, the cross-linking reaction starts, which after conclusion of the →*pot life*, leads to the hardening of the coating system. In this regard, the role of the hardener component can be different: in the case of isocyanate cross-linking systems (→ *polyisocyanates*) and two-pack epoxy systems (→ *epoxy resins*), the cross-linking takes place between reactive groups of the base material and those of the hardener component. This means that the hardener is chemically incorporated in the coating film; in the case of acid curing coatings (→ *acid curing*) and of the unsaturated polyesters (→ *polyester resins, unsaturated*), functional groups of the base material film former react with each other and the hardener serves only as initiator (→ *peroxides*) or as catalyst for the reaction.

two-pack systems: → *two-component systems*

U

ultra-filtration: Special technology employed in the electrodeposition coating, on one hand in order to continuously remove the low-molecular substances, which accumulate in the bath liquid, and on the other hand to provide sufficient amounts of ultra-filtrate for the rinsing of the uncoagulated coating adherent on the surface of the coated workpiece before stoving. Ultra-filtration takes place in filter modules by means of membrane filters, through which water and other low molecular substances are pressed while the high molecular polymer ions of the coating still remain in the module. The so produced ultra-filtrate is conducted through the different rinsing stages counter-current to the coated workpieces, so that these, during the rinsing process, are treated with increasingly cleaner ultra-filtrate. Subsequently, the ultra-filtrate as well as the concentrate generated by the ultra-filtration are led again to the bath.

ultramarine pigments: Aluminium silicates, which contain sulphur compounds as colour generating groups in their lattice structure gaps. Depending on the sulphur amount, they can be blue, green, red or violet. They are characterised by brilliant colours, high light fastness and temperature stability, that is why they are some time employed in the coil coating and powder coating segments. Conversely, their weather resistance in industrial atmosphere is so poor, that they cannot be used for outdoor-applications. Altogether, they play a secondary role in the coating technology. However, they are important for the production of printing inks and artist's colours.

ultraviolet radiation: Often simply called UV-radiation. Range of the electromagnetic radiation, which borders on the short wave limit of the visible spectrum. It comprises radiation with wavelengths from about

400 to 10nm and transmutes gradually into the Röntgen-radiation. Particularly interesting for the coating technology are the ranges of UV-A (between 400 and 320 nm) and UV-B (between 320 and 280 nm), which are relevant for the → *UV-curing*. But the UV-ranges with very short waves affect the coating industry as well, since they are responsible for the degradation processes in the coating films, caused by light.

underbody sealing: Application of a layer, which has to protect the underbody of the car against the high mechanical and chemical stresses which take place during driving. Among the various types rank in the first place the stone chipping, the steady abrasion caused by all kind of granular substances and the effects of water and salt. To this purpose, thick-film materials are employed, with film-thickness between 100 and 1000 µm, which show remarkable elastic properties and at the same time high abrasion resistance. As film formers, → *plastisols* and polyurethanes are predominantly used. The application of the underbody sealant normally takes place after coating with electrodeposition primer and seam sealing. In especially vulnerable areas the underbody sealant is often reinforced by an additional coating against stone chipping.

undercoat varnish: In the decorative paint technology usual pre-coating to improve the spreading rate of the topcoat. The undercoat varnish is cheaper than the topcoat and equipped with relatively much filler (extender), so that it is able to form an efficiently sealing layer on porous substrates like wood. Thus, the topcoat cannot be absorbed very much by the substrate and therefore can be spread on a larger surface area. This helps to realize relevant savings in the cost of material. Before application of the topcoat, the undercoat varnish has to be sanded.

undercuring: Cross-linking end condition of a two-component system, which is realised if too little an amount of hardener is added to the base material. Due to the lower cross-linking density, undercuring leads to reduced hardness and chemical and weather resistance properties in the coating system. Conversely, an increase in elasticity, and often also in adherence, takes place.

underwater coatings: Water resistant materials, which normally also have a seaweed repulsive function. They are mostly based upon high chlorinated film formers like chlorinated rubber and chloroprene, which are on the one hand resistant to hydrolysis and hardly susceptible to swelling by influence of water, while on the other hand they prevent or make more difficult the growth of algae, because of their chlorine content. This effect can be reinforced by the addition of biocides.

unsaturated fatty acids: → *fatty acids, unsaturated*

unsaturated polyesters: → *polyester resins, unsaturated*

urea resins: They belong, together with the melamine and benzoguanamine resins, to a group of polycondensation resins which are derivatives of carbonyl adducts (→*formaldeyde*), and they represent important reaction partners for the formulation of stoving and acid curing coating materials. However, due to their lower cross-linking density, they are not as weather and chemical resistant as melamine resins, and moreover at temperatures above 140 °C they have a tendency to yellowing. Because of their low price, they are of interest to application fields where the properties mentioned above are not important, for example common industrial coatings, industrial surfacers, and acid curing hammer finish paints.

urethane: Chemical compound, which is originated by the reaction between isocyanates and alcohols. Most reactions of isocyanate curing two-pack systems are carried out according to this mechanism. Due to their polymer nature, the cross-linked products are called → *polyurethanes (PUR)*.

UV-absorbers: Coating additives, which are able to absorb harmful UV-radiation and turn it into heat. Thus the coating films and/or the substrates are protected against premature degradation. Particularly important is the use of UV-absorbers in transparent wood coatings, where they have to prevent or retard the degradation of wood caused by light exposure, and furthermore in two-coat systems for the automotive coating, where they first of all prevent colour change due to pigment degradation, but at the some time they also protect the clear coat against yellowing and embrittlement. UV-absorbers display the best effect in combination with the so called "radical interceptors", which are able to convert already generated radicals (→ *radical*), as a consequence of film former degradation, into stable compounds and thus prevent them from continuing to participate in the degradation process.

UV-curing: It is the method of radiation curing (→ *electron beam curing*) most often used in industry. Although UV-curing does not reach the curing rapidity of electron beam curing (a few seconds against a split second), in most cases it is preferred because of the possibility to coat three-dimensional objects as well, and also because of its higher economy and the absence of dangerous gamma rays. The radical polymerisation or the cationic hardening process of the film former are started by means of highly energetic UV-radiation. However, in order to achieve a sufficiently quick radical formation it is necessary to employ special photoinitiators. The UV-curing process shows specific problems in the case of some pigmented systems (yellow, red and black), since UV-absorbing pigments change negatively the energy transfer and so impede an optimal through-curing. But even in such cases the employment of special UV-radiators allows to solve a lot of problems.

V

vacuum dissolver: Dispersing unit for the manufacturing of high viscosity coating materials, in the first place putties for the automotive refinish and for industrial applications. The use of vacuum in this case is necessary to avoid the inclusion of air blisters, which would cause an undesirable porous surface, during the sanding of the putty after its application and hardening. Modern vacuum dissolvers are equipped with a stripper blade, which periodically streaks over the inside of the dissolver vessel, in order to move over to the centre of the vessel the material fixed on the vessel side because of its high viscosity. Thus, it is insured that a homogeneous dispersion of the whole material takes place.

vacuum impregnation: Treatment method for wooden construction elements, which are exposed to biological degradation, mostly caused by weathering and some time additionally by contact with the ground. The vacuum impregnation makes possible the complete removal of air inclusions and of all absorbed gaseous components from the pores and fibres of the wood, so that the wood preservative, containing → *biocides*, can penetrate deeply into the wood structure and be effective against biological degradation.

value: Figure expressing a chemical or physical quantity, which serves to identify or characterise raw materials or products. Important chemical values for the coating technology are, for example, the acid value, the hydroxyl value and the epoxy value. Examples of physical values are the density and the refractive index.

Van der Waals forces: Also called dispersion forces. Physical forces of attraction, which interact between all kinds of atoms and molecules and are co-responsible for the cohesion of the matter (→ *dipole* and → *hydrogen bridge bond*). They are caused by temporary asymmetries in the charge distribution of the atoms involved. Van der Waals forces have shorter interaction reaches and are relevantly weaker than dipole forces and hydrogen bridges; they are the only active forces within non-polar compounds like for example → *aliphatic hydrocarbons*.

vapour cure process: In its modern version it is also called "vapour injection cure process". Process to accelerate the curing of two-component polyurethane systems without negatively influencing their →*pot life*. Shortly before application, a certain amount of a volatile amine, capable of activating a catalysts contained in the two-component material, is added to the atomisation air. Thus, the so activated catalyst accelerates the curing reaction only in the already applied coating film, without any negative influence on the processing time (pot life) of the cross-linked material.

ventilation: Technical equipment, which permits a controlled change of air in a working room. By the steady supply of fresh air and the simultaneous exhaust of polluted air, the ventilation equipment insures that the →*MWC-values* of hazardous substances are not exceeded and no explosive mixtures are generated.

ventilation surfacer: Special primer surfacer for wood, which permits a moisture exchange between the substrate and its environment. This is above all important for the coating of wood workpieces, which have not been stored optimally and therefore show too high a moisture content, or are steadily exposed to high humidity differences during their use. In these cases, to maintain the "respiration" of the wood is very important, otherwise the osmotic pressure arising under the coating layer would lead to flaking of the film. The pigmentation of the ventilation surfacers is, as a rule, close to the CPVC (→*critical pigment volume concentration*), some time even higher.

vibrating screen: Sieve variant, which is often used for purification of metallic coatings. It consists of a box shaped or circular frame in which a metal web with defined mesh size is fixed. The material to be sieved is filled in portions into the screen box. The box is connected with an eccentric, driven by a motor, which puts it into a vibrating movement. This avoids that the sieve meshes be obstructed by the aluminium or mica platelets.

vinyl acetate: →*polyvinyl acetate(PVAC)*

vinyl chloride: →*polyvinyl chloride (PVC)*

vinyl propionate: →*polyvinyl propionate*

viscometers: Devices for measuring the viscosity, that is the flow properties of liquid substances. In the coating technology, viscometers play a very important role, since the viscosity is one of the most relevant application properties of liquid as well as of powder coating materials (the latter have to be melted before measuring). The viscometers, which are usually employed for coatings, can be subdivided into two groups, according to their measuring principle: those where the shear (→*viscosity*) is caused by the effect of the force of gravity and those where it is produced by a mechanical rotation. To the first group belong the →*flow cup*, the capillary viscometer according to Ubbelohde and the falling-ball viscometer according to Höppler. However, an exact measure and evaluation of the flow properties of coating materials, showing marked →*viscosity anomalies*, is only possible by means of the different variants of →*rotation viscometers*, where the shear forces are mechanically produced and can be varied as one likes.

viscosity: It expresses which way a liquid is able to flow under the influence of mechanical forces. Scientifically it is called "dynamic viscosity" and, according to the Newtonian equation, it is defined as the shear stress divided by the shear rate. Here the shear stress expresses the force which must be employed to let flow (shear) two liquid layers, with a defined surface area, against each other at a certain velocity. If the ratio of the shear stress to the shear rate, at a given temperature, is constant, it is the case of an ideal or Newtonian flow behaviour. If this ratio changes with time and/or by variation of the shear stress, than we have a →*viscosity anomaly*. From the dynamic viscosity it is possible to calculate the "kinematic viscosity" by dividing it by the density of the liquid. This makes sense in the first place in the case of a viscosity measure by means of the capillary viscometer, because in this case the density of the liquid greatly influences the measuring results. Finally, it is absolutely important to consider that all variants of flow behaviour depend on the temperature. Thus, an exact viscosity evaluation is only possible after careful adjustment of the temperature of the liquid. In modern viscometers, this takes place by means of thermostats.

viscosity anomalies: Variants of the flow behaviour of liquids, which diverge from the Newtonian viscosity, that is those which depend not only on the temperature but also on the shear rate and/or on the shear time (→*Newtonian liquids* and →*viscosity*). For the coating technology the most important anomalies are the →*thixotropy* and the →*pseudoplasticity*.

visible light: Range of the electromagnetic radiation, which is able to activate the sensors in the human eye and thus to generate →*colour*. This range comprises radiation with wavelengths between 380 and 750 nm.

visual evaluation: Evaluation by means of the naked eye. It is now, as in the past, a decisive approval criterion for all coating tests, which deal with colour matching and the evaluation of coating surface quality. In the case of the comparison of metallic colours the impression one has under different visual angles is indispensable and cannot be replaced only by colour measuring. Surface quality criteria like orange peel

(→ *orange peel effect* or → *haze* can also be evaluated most reliably by the skilled human eye.

VOC-directive: VOC means "Volatile Organic Compounds" and is a measure of the potential pollution of the environment, which can be caused by the solvents contained in the coatings. This term was originally used in the Anglo-Saxon countries, first of all in the USA and Canada, in connection with the requirement of solvent reduction but it has now been adopted in the European Union as well. The VOC-directive provides for a gradual plan of solvent reduction, which will be applied step by step to all coating processing factories and workshops.

volatile matter: Total amount of the components in the coating formulation which, during the drying or the hardening, leave the coating material. As a rule it concerns organic solvents or water, but also reaction products may belong to it.

Vossen blue: → *Berlin blue*

W

wall paints: Collective term for all variants of dispersion paints, which are used to coat the interior and exterior walls of buildings (→ *wall paints, interior* and *exterior*).

wall paints, exterior: Architectural paints, based on water-thinnable polymer dispersions, which are used for the outdoor protection of buildings. As film formers styrene-acrylate copolymers are mostly employed, often supported by silicone resin emulsions to improve their weather resistance.

wall paints, interior: Coating materials based upon water-borne dispersions (→ *dispersion paints*), which are only suitable for indoor applications. In comparison to the → *wall paints, exterior* they are rather overcritically pigmented (→ *critical pigment volume concentration*) and therefore they have high water vapour permeability. On the other hand, the high pigment volume concentration makes possible the addition of large amounts of cheap white fillers (extenders) in order to partially replace the expensive titanium dioxide. Interior wall paints contain film formers (in the first place polyvinyl acetate and propionate) with lower light, weather and chalk stability than exterior wall paints.

wash primers: → *primers*

waste disposal: During the coating production and application, different kinds and amounts of waste may occur such as filtration residues, off quality batches, and overspray. The classification and handling of these types of waste is regulated by different national and international directives and laws. In Europe waste classification takes place according to the European Waste Catalogue (EWC). The national and international waste regulations are generally based on the same priority sequence. First there is the avoidance or reduction of waste by means of process improvement measures or replacement of raw materials in the coating formulations by more eco-friendly ones. Then follows the recycling of waste. Only if this option is exhausted, a thermal utilisation, that is for example → *special waste incineration*, is permitted. At the end of the waste disposal process is the landfill, although for most kinds of coating waste it can only be used after incineration, because they contain large amounts of organic compounds, capable of contaminating the ground water.

water-borne coatings: General term for coating materials, in which water is used as solvent and/or thinner instead of organic solvents. Exceptions should be considered the interior and exterior wall paints, based on aqueous dispersions, which normally are not called water-borne coatings (→ *dispersion paints*). Although water-borne coatings, depending on their components, may contain between 0 and 15 % organic solvents as → *co-solvents* or → *coalescence agents*, nevertheless they rank among the eco-friendly coating systems. Thus, in the automotive coating, in addition to the electrodeposition coatings, there are now also hydrosurfacers and water-borne base coats; in the industrial coating technology, water-borne stoving and two-component systems are increasingly employed; finally the architectural sector is more and more oriented to water thinnable decorative paints and wood glazes. Beside their compatibility with the environment and their biological harmlessness, the waterborne coatings show a high safety level concerning fire and explosion danger. Conversely, disadvantages are the high evaporation energy needed from water, which retards the drying process and increases the risk of sags and, in the case of metallic coatings, also of clouding, and the high polarity of water, which substantially impedes the wetting of some substrates, above all if the pre-treatment is insufficient. Also the dependence of the evaporation behaviour of water on the relative humidity in the air of the environs, which some times renders necessary the use of conditioned spray booths (→ *conditioned room*), has to be considered a negative. However, in most fields of application the positive aspects of the water-borne coatings prevail, so that a further expansion of this market segment can be expected in the future.

water, desalinated: It represents the second highest level of industrial water purification after the "distilled water". The latter is not only free of salts but also of microbes; however, because of its expensive production method, it is only used in the pharmaceutical industry. Water for general chemical processes is only desalinated by means of ion exchangers, that is it is free of foreign ions (→ *demineralized water*). This is also the case of water used as solvent in water-borne coatings, since foreign ions, above all calcium and magnesium deriving from the → *water hardness* but also iron and barium, would damage water dispersed film formers (→ *water solubility*). The consequence would be a coagulation of the formerly water soluble film former. Aqueous dispersions, which are predominantly manufactured according to the emulsifier-principle (→ *emulsion*) are, in relation to the tendency to coagulation, not so susceptible and are generally thinned by well or tap water; for example if they are used for wall paints, provided that they are supplied with softening agents.

water-free: Requirement referring to raw materials for the employment in isocyanate-based two-pack systems. Water is able to react with isocyanates under separation of carbon dioxide, which can provoke blistering and haze.

water glass paints: They are the only coating materials containing purely inorganic components as film formers. "Water glasses" are called the aqueous solutions of sodium and potassium salts of the ortho-silicic acid. However, in the coating technology potassium water glasses are employed almost exclusively. The solutions are very alkaline and therefore they show a clear affinity to silicate substrates which are alkaline as well, like concrete, cement, and mortar. Besides, the alkaline → *pH-value* affects positively corrosion protection. The film forming of such inorganic binders takes place in a first step by the formation of polysilicic acid and by the absorption of carbon dioxide, the latter reacting with potassium to form carbonate. In the course of time, the final → *silicification* of the polysilicic acid to silicone dioxide takes place with water separation. The so produced films are outstandingly resistant against mechanical stress and besides they show high light and → *weather resistance*, and furthermore good → *water vapour* permeability. The main application sectors are the restoration of old buildings and the heavy duty corrosion protection.

water hardness: It is the total amount of calcium and magnesium salts dissolved in a body of water. Internationally it is expressed in mmol per litre. One differentiates between "temporary" and "permanent water hardness". The temporary one consists of the dissolved amounts of calcium and magnesium hydrogen carbonates which, even by evaporation of only part of the water, precipitate as insoluble carbonates with separation of carbon dioxide (boiler scale). Conversely, permanent water hardness consists of easily soluble salts like sulphates or chlorides, which only after the almost complete evaporation of the water, that is by exceeding the saturation limit, form solid sedimentation. In the coating technology, water hardness is important, if water is used as solvent or thinner. In case of dissolved systems (→ *water solubility*) which are based on the mechanism of salt formation, the water employed has to be completely free of ions (→ *water, desalinated*). Otherwise, there is the danger of the formation of hardly soluble calcium and magnesium salts, which may lead to precipitation of the film former. Using water as a thinner (→ *water thinnability*), it is normally enough to employ a "soft water". That occurs above all in the production of dispersion paints. As a measure of precaution, these also contain in their formulation amounts of softening agents, which minimise the risk in case of thinning by the user.

water mountain: It describes the anomalous viscosity behaviour of waterborne coatings during their dilution with water. The salt groups, responsible for the → *water solubility* of the binder, form by dissociation electrically charged particles (ions), which have a pronounced tendency to association with the → *dipole molecules* of the water by means of electrostatic forces of attraction. The consequence of it is that, during the addition of the first water portions, the binder molecules generate associates among each other by means of water, which have lower mobility than the single molecules. Thus, the viscosity of the coating system increases rather than decreasing, as it would be normally expected by dilution. This phenomenon continues, until a maximum viscosity value is reached, called "water mountain". The further addition of water portions separates more and more the associate molecules, until their forces of attraction break down and the viscosity decreases rapidly. In practice, the water mountain represents a very negative property, because it interferes with or makes it impossible to exactly adjust the application viscosity. Only the simultaneous use of organic solvents, so called co-solvents, makes it possible to level the water mountain and so to solve this problem.

water repellent (hydrophobic): Property, which is mostly achieved by the employment of silicone coatings or of silicone modified systems. The water repellent effect is caused by both the low polarity of the polysiloxanes as well as by the high polarity of the water. Since here no wetting of the coated surface takes place (the water drips off), the binder is protected against degradation processes caused by hydrolysis and as a consequence it shows pronounced weather resistance. That can be seen, for example, by the application of silicone polyesters in coil coating materials for façade components or in high performance, silicone modified exterior wall paints. However, the high price of the silicone components is a disadvantage. Another very up-to-date possibility to cause water repellence is the formation of hydrophobic nano-structures on the surface of the coating, which generate a so called "lotus-effect" (→ *nano-technology*).

water resistance: Important quality feature for coating materials and systems for the outdoor application as well as for others which, during their use, often come into contact with water and/or water solutions. A precondition for good water resistance is that the coating components show only low hydrophilic properties and the film former has a pronounced resistance against hydrolysis, that is separation reactions by means of water.

water solubility: Property of many binders, which are employed in water-borne systems. Water solubility leads in the water medium to the formation of true or colloidal solutions (→ *solution, true* and *solution, colloidal*) and has to be distinguished from the mere → *water thinnability*. The precondition for the water solubility of organic film formers, which have rather non-polar structures, is to equip them with a sufficiently high amount of electrical charges, able to procure for the dipole water the necessary points of anchoring, in order to form a stable → *solvation wrapping*. That takes place by means of a salt formation mechanism, where mostly acid, only poorly dissociated carboxyl groups of the film former (only in case of cathodic electrodeposition binders one has basic amino groups) are turned into completely dissociated salt groups by neutralisation. Caused by electrolytic dissociation of these salt groups, ions are generated, that is electrically charged particles, which temporarily make possible the anchoring of water molecules and thus water solubility. As a consequence of evaporation of the water as well as of the neutralisation agent, generally amines, the process of salt formation is reversed so that water solubility is lost again. Therefore, during the processing of water-borne coatings, major attention has to be paid not to permit the formation of dried coating remains in vessels and pipelines as well as in dispersing and application devices, which would inevitably lead to speck formation, since the water is not capable to dissolve them any more.

water thinnability: Inherent property of all water-borne coating materials. However, it differs from the → *water solubility,* since not all water thinnable coatings and paints must necessarily contain water soluble binders. In the case of aqueous dispersions for interior and exterior wall paints, for example, it is not a matter of true solutions of the binder (→ *dispersion* and *solution, true*), since the binder particles have a diameter above 0,1μm. Their "ability to float" is mostly achieved by the fact, that the particles are wrapped in hydrophilic emulsifier layers, which are stabilised by so called protective colloids. Also such coarsely dispersed aqueous systems can be thinned by water like the water soluble ones. Besides, they have the advantage, during the thinning process, to show no inclination to the → *water mountain* phenomenon.

water treatment: This term refers to technologies for the production of suitable process water, that is the softening (→ *water hardness*), the desalination (→ *water, desalinated*) and the water distillation as well as to measures of treatment of sewage before its introduction into rivers and lakes or into the sewers.

water vapour permeability: Primary property of coats, which are generated by dispersion paints, above all those for the indoor application (→ *wall paints, interior*). It is evaluated according to DIN 52 615 and it expresses the ability of wall paint layers to deliver into the atmosphere the water vapour absorbed by capillary effects, in case of a sudden change in temperature and humidity, without that the generated osmotic pressure leads to flaking or blistering. Since such sudden changes take place above all indoor, caused by heating, cooking and washing processes, interior wall paints have to be particularly "able to breath" . Thus they are mostly over-critically pigmented (→ *critical pigment volume concentration/CPVC*).

wavelength: Fundamental value for the definition of the properties of electromagnetic radiation, to which belong the ranges of the UV-radiation, of the visible spectrum and of the IR-radiation, important for the coating technology (→ *ultraviolet radiation, visible light* and *infrared drying*). Electromagnetic radiation diffuses in the form of waves, according to a sine function. The wavelength λ expresses the distance between two equal points of two successive waves, for example between two maxima or two minima of the waves. Since the velocity of propagation of the radiation results from the wavelength multiplied by the oscillation frequency (number of oscillations per time unit), the frequency of a radiation with constant velocity is the higher the smaller its wavelength. On the other hand, the frequency is a measure of the energy content of the radiation; the higher the frequency the higher the energy and also the destructive potential of the radiation. Thus, it is obvious why the short wave UV-radiation shows a more damaging effect, for example during a weathering process, than the longer wave visible light.

waxes: Term for a group of natural and synthetic products, which at certain temperatures are kneadable and able to polish and in the melting state show low viscosity. The natural waxes are esters of high alcohols and long-chain fatty acids. Among the synthetic waxes, paraffin waxes are the best known. They consist of blends of solid, aliphatic hydrocarbons, which are produced by mineral oil distillation. In the coating technology, beside polyethylene and polypropylene waxes, PTFE and polyamide waxes are mostly used. They are preponderantly employed as slip and anti-blocking additives but also as flatting agents.

weathering, accelerated: → *weathering, artificial*

weathering, artificial: This term refers to several short test methods, which serve to simulate different climate conditions, like for example → *weather-o-meter*, sunshine-test, and QUV. These tests cannot replace the outdoor weathering, because a 100 % correlation is not possible; however they are indispensable, if one needs quick information about the general weather and light resistance of coatings. This necessity occurs above all in the case of the development of new products or of the comparison between materials of different competing producers

weathering station: Place with suitable technical equipment, in order to carry out outdoor weathering tests of coating materials and coating systems under desired natural conditions. Some coating companies and coating raw material manufacturers dispose of own weathering stations. However, in most cases such facilities are provided commercially, since the required climate conditions can only be realised in far away regions (→ *Florida test*). The panels to be weathered are sent there by the coating companies and are exposed by the station operators. During the weathering, evaluations take place periodically, whose results are communicated to the customer in the form of reports. Important weathering stations exist for example in the Alps region, in different European sea and industrial climate zones, in Florida, Arizona and Australia.

weather-o-meter: Device for accelerated weathering to simulate degradation processes in coatings, which are exposed to humid/hot climates with high UV-radiation (→ *weathering, artificial*). The needed UV-radiation (→ *ultraviolet radiation*) is generated by xenon-, fluorescence-, halogen- or more seldom carbon arc lamps. Suitable filters provide that only the UV-radiation contained in the day light reaches the test panels.

Around the lamp a cylinder rotates, at which the coated panels to be tested are fixed. During the exposure, radiation phases alternate with dark phases equivalent to the day-night cycle. At intervals irrigation with desalinated water takes place. The xenon-test works similarly to the weather-o-meter. In the coating technology both are considered standard methods for a quick anticipation of the time consuming → *Florida test*, even if their results do not exactly correspond.

weather resistance: Stability of a coating against effects, caused by weather factors like moisture, heat and light, in some cases also combined with atmospheric polluting components like acids and salts. Since sun radiation represents an important climatic factor, the weather resistance tests include automatically an indirect evaluation of the → *light stability* of the components, which are contained in the coating material. To test weather resistance, different variants of → *outdoor weathering* in different climate zones are available as well as numerous short tests (→ *weathering, artificial*).

wet abrasion resistance: Important property of dispersion paints, above all such for the indoor application, which are some time cleaned by scrubbing. The washing and scrubbing resistance is evaluated according to DIN 53778 (DIN ISO 11998) with the Gardner scrubbing tester (→ *wet-rub resistance*). Here, two brushes or other scrubbing materials move forwards and backwards on the paint layer to be tested with supply of scrubbing liquid. The evaluation takes place either after a certain number of scrubbing cycles or when a defined damage pattern has been formed.

wet film thickness: Thickness of a coating film immediately after its application. It can be evaluated in a simple way by means of the comb gauge according to Rossmann or of the rolling rim (also "twin wheel" or "measuring wheel"). In the first case a metal comb, consisting of two feet with equal length between which teeth of varying length are located, is put perpendicularly onto the film surface. The first tooth, which is wetted, indicates the wet film thickness. The rolling rim consists of three wheels, whose middle one is slightly eccentric. The rolling rim is rolled along the wet film until the middle wheel just touches the film surface. The corresponding film thickness can be directly read from a scale at the side. The measuring of the wet film thickness is above all important as process control method in industrial coating processes.

wet-on-wet-application: Process for certain multilayer-systems, where a coating material, for example a surfacer, after its application is only briefly physically dried, before being overcoated by the next coating material, as a rule a top-coat. This process is particularly used for two-coat systems in the automotive area, where a base-coat layer, which can only dry physically, after a short flash off is oversprayed with the protecting clear coat (topcoat). In the case of heavy vehicles and industrial coating, wet-on-wet-surfacers are often employed, whose final hardening takes place after overcoating.

wet rinsing systems: Current technology for the efficient removal of spray dust from the spray booth (→ *overspray*). Wet rinsing systems are above all employed in industrial and manual spray application processes, where the overspray absorption takes place below the booth floor by means of circulating water. The air stream in the booth is conducted downwards (→ *spray booth*) and it transports the spray dust, generated by the spray application, to the water surface. There the dust is wetted by the flowing water and removed from the spray booth. The regeneration of the water, or alternatively its treatment before disposal, takes place by the addition of special coagulants, which leads to the coagulation of the coating droplets dispersed in the water. Thus, these sink to the bottom and form there a so called "coating sludge", which periodically can be removed as special waste or, on certain conditions, can also be reused. In the past, laboratory spray booths or spraying stations also often employed wet rinsing which now has been replaced almost completely by filter mats, because here the water treatment is very problematic.

wet-rub resistance: Important technological property of dispersion based interior wall paints. Since such paints, as a rule, have an overcritical pigmentation (→ *critical pigment volume concentration*) and therefore are mat, relatively porous and thus susceptible to dirtiness, their ability to cleaning by rubbing with wet clothes is of much practical importance. Doing so, dirt traces have to be removed without rubbing off the dispersion paint itself. Wet-rub resistance is normally evaluated according to DIN 53778/part 2 (→ *wet abrasion resistance*).

wet sanding: Pre-treatment method for coating layers, which have been applied on substrates and are already dried, necessary before their overcoating. The wet sanding makes possible, on the one hand the dust-free smoothing of uneven spots in the old coating layer, and on the other hand the increasing of the contact surface between the old and the new layer and so an improvement of the adhesion. Wet sanding is not suitable for absorbent substrates like wood and pasteboard, which would be swollen by the sanding water (→ *dry sanding*).

wet sieving: Method for separation of particle heaps into different grain fractions, which is used above all in the production process of fillers (extenders) from minerals and for the analytical evaluation of their particle size distribution. The extender, in the form of an aqueous suspension, is given onto sieves with different mesh sizes and thus separated in fractions with different particle sizes. After their drying, the grain fractions can be weighed, in order to determine the particle size distribution, or directly employed in a selective way. However, as an analytical method wet sieving is more and more replaced by electronically assisted methods like Coulter Counter or laser light diffraction (→ *particle size distribution*).

wetting: Process, during which a liquid spreads spontaneously on the surface of a solid and adheres on it durably. In the coating technology good wetting is very important relative to the → *dispersion process* of the pigments as well as to the adherence of the coating layer on the substrate after application. The precondition of good wetting is, that the → *surface tension* of the liquid should not be higher than that of the solid substance. However, particularly when referring to the coating of plastics and the dispersion of pigments, the situation concerning the surface tensions of the binder/solvent-mixture and of the solid surface to be wetted is not ideal, the wetting process has to be supported by the addition of suitable wetting agents.

wetting agents: Also called dispersing additives. They rank among the most important coating additives, since without them pigment dispersion in the coating material scarcely succeeds (→ *wetting* and *dispersion process*). Their duty is to balance the surface tension differences between film former solution and pigment surface, because this is the precondition for good wetting. Besides, they contribute also to the wetting of the substrate. In order to fulfil this purpose, wetting agents have to contain in their molecules chemical groups which are "pigment-friendly" as well as those which are "binder-friendly". With the pigment-friendly groups they anchor on the pigment surface, while with the binder-friendly ones they form a wrapping around the pigment particle, which is orientated outwards to the film former and is able to be wetted. (Concerning the subdivision of the coating agents due to their chemical structure → *ionic* and *non-ionic*).

white goods: Familiar term for household appliances like refrigerators, washing-machines and dish-washers. The coating of their components, which are mostly phosphated or pre-treated with primer by means of electrodeposition, takes place by → *electrostatic application*, whereby increasingly waterborne coatings and powder coatings are employed. For the less visible lateral parts, many manufacturers use also → *coil coating* sheet metal.

whitening power: Characteristic property of white pigments. It expresses their ability to produce more or less bright colours in a mixture with coloured or black pigments. The higher the whitening power of the white pigment, the brighter are mixtures containing equal pigment ratios.

white pigments: Due to its superior refractive index, nowadays almost exclusively → *titanium dioxide* is employed. Other pigments, which during the centuries had been very important, are now used only infrequently. White lead, an easily soluble basic lead carbonate, is forbidden for coating applications because of its toxicity. Zinc oxide is just only employed as a modification component for → *zinc phosphate* in the corrosion protection. Only → *zinc sulphide* and → *lithopone* are sporadically used in marginal sectors.

white rust: → *moisture-induced corrosion*

white spirits: Term for a group of aliphatic solvent blends derived from petroleum distillation, with a flash point above 21 °C and defined boiling limits. As a rule, they also contain little amounts of aromates. According to their boiling limits, white spirits are subdivided in a ranges of 1 to 5. White spirits represent the most non-polar solvent group and are the main solvents for decorative paints and → *cyclised rubber paints*.

white standard: Standard with known lightness value (Y-value), suitable to calibration of spectrophotometers (→ *colour measurement instruments*). In the case of old devices, for this purpose it was necessary (according to a DIN standard) to make a tablet of barium sulphate with a prescribed degree of purity. Today, only adequate standards of ceramics are used.

wood coatings: General term for all coating materials, which are applied onto wood substrates. It includes wood putties, wood primers and surfacers (→ *ventilation surfacer*), undercoat varnishes, top and clear coats. A special kind of top coats are the → *wood glazes*, which have to fulfil special protection functions. Referring to the application, one has to differentiate between wood coatings for the manual processing (→ *decorative paints*) and those for the industrial application. In the first case, mostly oxidative drying alkyd resins, nitrocellulose paints and water-borne dispersions are employed. In the industrial sector, mostly two-pack polyurethane coatings, unsaturated polyesters and acrylic resins for the UV-curing but also waterborne impregnating agents and glazes are preferred.

wood glazes: Wood coatings, equipped with transparent or semi-hiding pigments or dye stuffs, which let shine through the grain and structure of the substrate. They are often coloured with transparent iron oxide pigments and carbon black, since these, similar to the wood stains, permit the imitation of different wood types. Moreover, due to their high absorption in the UV-range, they protect the wood against aging caused by light. Today, also wood glazes equipped with other coloured or white pigments are increasingly popular. Wood glazes have very low viscosity, since on the one hand they have to penetrate as deeply as possible into the wood fibres in order to guarantee a lasting protection, and on the other hand they must produce open pore coating layers able to later permit a self-controlled regulation of the wood moisture. To improve protection against UV-radiation, wood glazes are as a rule additionally equipped with → *UV-absorbers.* They also contain biozides, which prevent or retard the wood degradation by micro-organisms. In solvent-borne wood glazes, oxidative drying alkyd resins are employed as film formers. The water-borne variants are based on acrylic-dispersions.

wood moisture: Natural component of all wood-based materials. After a sufficient time of storage of the wood before its processing, the moisture reaches a natural balance which, according to the wood type, can fluctuate between 8 and 15 %. The keeping of these moisture values has a major importance for the coating of precision components of solid wood, because a subsequent loss or absorption of moisture would lead to a change of shape (warping of the wood). In those cases, the elasticity of the coating film may not be able to compensate for the deformation. Besides, coating application on too moist wood could provoke blistering and flaking.

wood oil: Oxidative drying oil of vegetable nature with outstanding drying properties. These are caused by the eleaostearic acid, containing in its molecule three double bonds which, due to their position, can react particularly quickly. In the past, wood oil was often converted to oil varnishes with phenolic resins. Today it is exclusively used to produce drying alkyd resins, whereas as a rule it is added as a component to mixed alkyds to improve their drying properties. Pure wood oil alkyd resins are seldom used, because of their strong tendency to skin formation and gelling.

wood rosin: Rather dark variant of the → *colophony,* which is extracted from the roots of the conifers. Before use the wood rosin is mostly brightened by bleaching.

wood stain: → *staining*

wool scale: Measure for the evaluation of the fading or the darkening of colours as a consequence of light exposure. It is described in the DIN 54 003 and it serves to compare the test panel with differently coloured wool threads under a defined illuminant. While the wool scale in the textile industry still conserves a certain importance, in the coating technology it is considered antiquated and is almost completely replaced by colorimetric methods.

wrinkle finishes: In the past very popular, however nowadays hardly employed, mostly black structured paints for machines and office equipment (for example typewriters). They are based upon oxidative drying alkyd resins, which are deliberately equipped with an excessive amount of cobalt drier. Therefore, surface drying take place very quickly, while depth drying progresses only slowly. Caused by the enclosed solvents, the inferior coating layers remain liquid for a long time, so that the dried skin practically swims on them. Since the latter has a larger volume than the non hardened coating layers, due to the oxygen absorption, it gradually forms little wrinkles.

wrinkling: It may be caused by formulation as well as by application mistakes. Some time there are also formulations which aim deliberately at wrinkling (→ *wrinkle finishes*). Typical formulation mistakes are the excessive addition of driers to oxidative drying coatings or too many low boilers and/or too aggressive solvents in stoving coatings. Examples of application mistakes are the too quick heating of wet stoving coating films, which may impede the controlled volatilisation of the solvents coming out from the lower layers, and the overcoating of coating layers which are sensitive to solvents (→ *lifting*).

X

xenon test: → *weather-o-meter*

xylene: It is the most important solvent for coatings among the aromatic hydrocarbons. It occurs as a technical blend of three isomers (ortho-, meta- and para-xylene) with different amounts of ethyl benzene and in its dissolving power it can be compared with → *toluene*. Its broad suitability ranging from non-polar to medium-polar coating formulations, its advantageous evaporation behaviour in addition to a flash point above 21 °C and finally its low price combined to make xylene the most important solvent until a few years ago. However, since its → *MWC-value* has been clearly reduced because new medical tests results advise it, requirements for its replacement by less dangerous medium boiling solvents are increasingly proposed. In relation to this change, the importance of → *butyl acetate* has largely increased. However, xylene is still one of the more often employed solvents.

xylene/formaldehyde resins: Also called XF-resins. They are produced by addition of formaldehyde to xylene and subsequent condensation of the so formed methylol groups (→*formaldehyde*). The XF-resins cannot be further cross-linked and are used as non-saponifiable polymer plasticisers (→*soft resins*) in physically drying coating systems.

Y

yellowing: Phenomenon, which is attributed to alterations in the binder matrix of a coating material. In some variants of binders the yellowing has to do with the formation of colouring groups in their molecules, as a consequence of reactions caused by aging, which are not necessarily connected with degradation processes, for example in alkyd or phenolic resins. However, in other binder systems the yellowing gives the early information about film former degradation caused by weather influences. In this connection the effect of highly energetic radiation (→*ultraviolet radiation*) is particularly important. As a rule, the yellowing represents (in bright colours) the first visible stage of degradation phenomena caused by the weather, followed by the loss of gloss (→*gloss retention*) and finally by the →*cracking*, which practically has to be considered the final stage.

yield: →*spreading rate*

yield point: Characteristic value to determine the right pigmentation ratio for mill bases containing inorganic pigments and/or fillers (extenders). The evaluation of the yield point (Daniel-titration) is simultaneously connected with the determination of the “smear point”. A defined amount of pigment or filler is weighed into a beaker. Under steady stirring with a glass bar, first a sufficient amount of binder solution is added to produce a paste, which is just able to smear the glass (smear point). If enough binder solution is then added, that the obtained paste just flows down from the glass bar, the “yield point” is reached. The smear point gives an indication relating to mill basis formulations for triple roll mills and dissolvers, while the yield point is rather suitable for bead and ball mills.

Z

zapon lacquers: Term referring to physically drying, clear or transparent coloured coatings with a low solids content, mostly based on →*cellu-*

lose nitrate. They offer poor surface protection, but they dry extremely quickly and they can also be formulated to produce highly elastic films. They are only used as one-layer-coatings and their main fields of application are the coating of wood, bare metal, glass and leather. Here they often combine the mechanical protection with a decorative function.

zeolites: Hydrous silicate minerals, whose crystal structure is pierced by cavities and channels. By heat effect, the water of crystallisation can be separated reversibly. Water free zeolites show a pronounced tendency to water absorption and besides they are able to capture into their hollow spaces the little water molecules. Therefore they are used in the coating technology as so called "molecular sieves", in order to deprive of moisture water sensitive systems like, for example, isocyanate solutions. Special types are also employed for the reversible absorption of solvents or pollutants.

zeta potential: It is the electrical charge of a pigment surface, or of a stable layer adsorbed on it, which takes effect outwards. In the case of a dispersed pigment, the adsorbed layer can consists of solvent and/or wetting agent molecules. The zeta potential can be considerably influenced by post-treatment of the pigment after its production. This fact is largely utilised by the pigment manufacturers to improve the pigment dispersibility in different media. In water borne systems, the zeta potential of the pigment also depends on the →*pH value.* Depending on the system composition, a pH value exists at which the surface charge is zero, a state known as the "isoelectric point (IEP)". The influence on the zeta potential by variation of the pH value is used, among other things, to stabilise pigment dispersions by electrostatic repulsion.

zinc (to): Also commonly called galvanising. Post-treatment process for steel to improve its corrosion protection. There are two different variants: hot dip galvanising and electro-galvanising. In the case of the →*hot dip galvanising* the steel workpiece, for example the sheet steel, is dipped into a zinc melt. Thus, zinc layers are formed, which can have a film thickness up to 80µm and show a characteristic pattern. A special variant of the hot dip galvanising is the Sendzimir-process, where the sheet steel is conducted through the zinc melt and after that rolled to produce particularly thin layers with a fine pattern. Hot dip galvanised steel substrates are problematic for coating processes and need a pre-treatment with an adhesion promoting →*primer.* Besides, the zinc layer is often porous or containing air inclusions. That implies the danger of blistering in stoving coatings. Conversely, the electro-galvanising does not produce any pattern but an uniform, rough, grey zinc layer, which normally can be coated without problems. The corrosion prevention mechanism is very effective and comparable to that of the → *zinc dust primer.*

zinc chromate: In the past the most employed anticorrosive pigment for industrial and automotive applications (→ *corrosion protection OEM, refinish, industrial*). Its outstanding anticorrosive effect was based on the quick formation of passivating layers, due to an adequate water solubility. Because of its carcinogenic potential, nowadays it is used much less (→ *chromate free*). Instead → *zinc phosphate* is now mostly employed in the industry as replacement pigment.

zinc dust: It is the most important anticorrosive pigment for heavy duty corrosion protection (→ *corrosion protection, heavy duty*). Zinc dust pigments for the coating technology consist of almost spherical particles, with an average particle size between 1 and 10μm, depending on the type, and have a zinc content of at least 98 %. The corrosion protection effect is caused by the fact that zinc is more non-noble than iron (→ *electrochemical series*) and therefore, under the influence of corrosion agents, it behaves as a "sacrificial anode" and it dissolves instead of iron. The oxidation products caused by reaction with the air oxygen and water, differently from iron rust are well adhering and compact and form an efficient barrier which prevents further corrosion aggression. The fact that these compounds are also very voluminous allows them to cement scratches and crevices produced by defects or damages of the zinc layer. The same corrosion prevention mechanism takes place in the case of galvanised steel (→ *zinc*) as well.

zinc dust primers: They are mostly high-build-systems for the heavy duty corrosion protection, which normally contain only zinc dust as a pigment. Their pigmentation ranges between 70 and 90 %. That means that, in spite of the very high density of the zinc dust, the pigment volume concentration (PVC) exceeds the critical value (→ *critical pigment volume concentration/CPVC*). Different from the other corrosion protection systems, where the PVC-value has to be significantly lower than the CPVC, the zinc dust primers need an over critical pigmentation, in order to develop their full effect. This depends on the fact that the anodic protection of the iron, by means of the zinc, only takes place if the two metals contact each other. A too low PVC means that the zinc dust particles are completely wrapped by the binder, so that no electrolytic interchange with the iron could take place. As film formers for zinc dust primers are mostly employed two-componentsystems (→ *polyisocyanates* and → *epoxy resins*) or → *silicic acid esters.*

zinc oxide: Also known as “zinc white”, it represents, together with white lead, one of the oldest white pigments. However, since the appearance of → *titanium dioxide* it has lost almost completely its importance as a white pigment for the coating technology, because of its relatively low refractive index. After the widespread acceptance of → *zinc phosphate,* zinc oxide has experienced a certain revival as auxiliary pigment in corrosion protection. In this regard, zinc oxide improves the initial corrosion protection of zinc phosphate by neutralisation of acid corrosion initiators, due to its basic properties. To this purpose it is added in little amounts to the formulations of primers or it is directly included into the phosphate pigment as modification component.

zinc phosphate: → *phosphate pigments*

zinc phosphating: → *phosphating*

zinc sulphide: After titanium dioxide it is the white pigment with the highest → *whitening power,* although its refractive index is clearly inferior to that of the rutile (2.37 versus 2.75). However, different from the other white pigments, which in the coating technology are almost completely replaced by titanium dioxide, zinc sulphide has maintained some useful application segments. Above all as the whitening component of the → *lithopone* it is employed in putties and highly filled wall and floor paints. Because of its lower absorption capacity in the UV-range it lends itself to replace titanium dioxide in some UV-radiation-curing coatings.

zinc tetraoxichromate: Until recently it was the classical anticorrosive pigment for all → *wash primers,* since it was able to achieve a complex chemical reaction with the phosphoric acids and the metallic substrate, which warranted beside very effective corrosion prevention in addition to outstanding cohesion. However, because of its carcinogenic effect (→ *chromate-free*) today it is almost completely replaced by zinc phosphate.

zinc white: → *zinc oxide*

zirconium drier: In the modern formulations of oxidative drying coating systems it is employed as a replacement for lead drier, as a rule in combination with cobalt (→ *lead-free* and *driers*), because zirconium compounds do not have any substantial toxicity. In this combination a division of work takes place, so that the cobalt drier promotes more the surface hardening and the zirconium drier, similar to lead containing types, more the through hardening.

A

abrasion resistance: Abriebbeständigkeit
accelerators: Beschleuniger acceptance test: Eingangskontrolle
achromatic colour: unbunte Farbe acid curing: Säurehärtung
acid curing coating: SH-Lack
acid resistance: Säurebeständigkeit
acids: Säuren
acid value: Säurezahl
acrylic resins: Acrylatharze
addition polymerisation: Additionsreaktion
additive: Additiv
additive colour mixture: additive Farbmischung
adhesion: Adhäsion
adhesion primer: Haftprimer
adhesion tests: Haftungsprüfung
adhesive strength: Haftung
aeration: Belüftung
aerosol: Aerosol
aerosol lacquers: Sprühdosen-Lacke
after-treatment: Nachbehandlung
agglomerates: Agglomerate
aggregates: Aggregate
agitator: Dissolver
airbrush technique: Airbrush-Technik
airless spraying: Airless-Spritzen
airmix-spraying: Airmix-Spritzen
air separation: Windsichten
alcohols: Alkohole
aliphatic hydrocarbons: aliphatische Kohlenwasserstoffe
alkali resistance: Alkalibeständigkeit
alkyd resins: Alkydharze
alkyd resins, long oil: Alkydharze, langölig
alkyd resins, medium oil: Alkydharze, mittelölig
alkyd resins, short oil: Alkydharze, kurzölig
alternating atmosphere test: Wechselklimatest
aluminium: Aluminium
aluminium bronzes: Aluminiumbronzen
amines: Amine

amino resins: Aminoharze
anatase: Anatas
angle of incidence: Einfallswinkel
anion: Anion
anode: Anode
anti-corrosive pigments: Korrosionsschutzpigmente
anti-floating agents: Antiausschwimmmittel
anti-settling agents: Antiabsetzmittel
anti-skinning agents: Hautverhinderungsmittel
anti-static agent: Antistatikum
anti-static cloth: Staubbindetuch
application methods: Applikationsverfahren
application technological properties: anwendungstechnische Eigenschaften
application viscosity: Verarbeitungsviskosität
aqueous dispersion: wässrige Dispersion
architectural paints: Bautenlacke
areometer: Aräometer
aromatic-free: aromatenfrei
aromatic hydrocarbons: Aromaten
atomisation: Zerstäubung
attrition mills: Rührwerksmühlen
automotive coating: PKW-Lackierung
automotive (OEM) fillers: Autoserienfüller
automotive (OEM) primers: Autoseriengrundierungen
automotive (OEM) topcoats: Autoseriendecklacke
automotive refinishing coatings: Autoreparaturlacke
automotive refinishing process: Autoreparaturlackierung
automotive refinishing systems: Autoreparatursysteme
auxiliary solvents: Colöser
auxiliary products: Additive
azo-pigments: Azo-Pigmente

B

ball mill: Kugelmühle
barium sulphate: Bariumsulfat
barrel coating: Trommellackierverfahren
barrier effect: Barrierewirkung

baryte: Schwerspat
base material: Stammlack
bases: Basen
bead mills: Perlmühlen
Bénard's cells: Bénard'sche Zellen
bentonite: Bentonit
bentonite paste: Bentonit-Paste
Berlin blue: Berliner Blau
binder requirement: Bindemittelbedarf
binders: Bindemittel
biocides: Biozide
black pigments: Schwarzpigmente
blanc fixe: Blanc-fixe
blazing of plastic surfaces: Flämmen von Kunststoff
bleeding: Ausbluten
blistering: Blasenbildung
blocked, chemically: verkappt, chemisch
blocked isocyanates: blockiertes Isocyanat
blocking resistance: Blockfestigkeit
blown oil: geblasenes Öl
blue stain inhibition: Bläueschutz
blue-yellow value: Blau-Gelb-Wert
blushing: Weißanlaufen
boiled oil: Firnis
boiling: Verkochung
boiling temperature: Siedetemperatur
breathing protection: Atemschutz
bronzing: Broncieren
brush: Pinsel
brushing: Streichen
bulking value: Schüttvolumen
butanol: Butanol
butyl acetate: Butylacetat
butyl glycol: Butylglykol

C

cadmium-free: cadmiumfrei
cadmium pigments: Cadmiumpigmente

calcite: Calcit
calcium carbonate: Calciumcarbonat
camouflage paints: Tarnfarben
can coating lacquers: Doseninnenschutzlacke
carbon black: Ruß
carboxyl group: Carboxylgruppe
castor oil: Ricinusöl
castor oil, dehydrated: Ricinenöl
catalyst primer: Reaktionsgrund
catalysts: Katalysatoren
cathode: Kathode
cathodic electrodeposition coating: Elektrotauchlackierung, kathodisch
cation: Kation
cellulose acetobutyrate (CAB): Celluloseacetobutyrat (CAB)
cellulose nitrate: Cellulosenitrat
cement resistance: Zementechtheit
chalk: Kreide
chalking: Kreidung
channel carbon black: Gasruß
chemical resistance: Chemikalienbeständigkeit
china clay: China Clay
chlorinated hydrocarbons: Chlorkohlenwasserstoffe
chlorinated paraffins: Chlorparaffine
chlorinated rubber: Chlorkautschuk
chromate-free: chromatfrei
chromaticity diagram: Normfarbtafel
chromating: chromatieren
chromium oxide green: Chromoxidgrün
chromium titanium yellow: Chromtitangelb
CIE-L*a*b* system: CIE-L*a*b*-System
circulating air oven: Konvektionsofen
citric acid esters: Citronensäureester
clear coats: Klarlacke
closed pore: geschlossenporig
clouding: Wolkigkeit
coagulation (of an emulsion): Koagulation (einer Emulsion)
coalescence agents: Koaleszenzmittel
coating formulation: Lackrezeptur
coating materials: Beschichtungsstoffe

coating of plastics: Kunststofflackierung
coating production: Lackherstellung
coatings for plastics: Kunststofflacke
coating systems: Beschichtungssysteme
cobalt blue: Kobaltblau
cobalt drier: Kobalttrockenstoff
coconut oil: Kokosöl
cohesion: Kohäsion
coil coating: Coil-Coating
cold check test: Kälteprüfung (von Holzlacken)
cold resistance: Kältebeständigkeit
collodion cotton: Kollodiumwolle
colloidal solution: kolloidale Lösung
colophony: Kolophonium
colour: Farbe
colour, achromatic: Farbe, unbunte
colour, chromatic: Farbe, bunte
colour circle: Farbkreis
colour contrast: Farbkontrast
colour difference: Farbabstand
coloured pigments: Buntpigmente
colour fastness: Farbtonbeständigkeit
colour formulation calculation: Farbrezepturberechnung
colour index (C. I.): Colour-Index (C. I.)
colour location: Farbort
colour measurement: Farbmetrik
colour measurement instrument: Farbmessgerät
colour standard: Farbstandard
colour value: Farbzahl colourant: Farbmittel
combustibility: Brennbarkeit
combustion: Verbrennung
compatibility: Verträglichkeit
complementary colours: Komplementärfarben
compound (chemical): Verbindung (chemische)
compressed-air spraying: Druckluftspritzen
condensation resins: Kondensationsharze
conditioned room: Klimaraum
contact angle measuring: Randwinkelmessung
convection drier: Konvektionsofen

cooling by evaporation: Verdunstungskälte
copolymerisation: Copolymerisation
copper phthalocyanines: Kupferphthalocyanine
corona charge: Korona-Aufladung
corona discharge: Korona-Entladung
corrosion: Korrosion
corrosion protection, heavy duty: Korrosionsschutz, schwer
corrosion protection, OEM, refinish, industrial etc.: Korrosionsschutz, leicht
corrosion protection primers: Korrosionsschutzgrundierungen
corrosion tests: Korrosionsschutzprüfungen
co-solvents: Colöser
cracking: Rissbildung
crackling finish: Reißlack
cratering: Kraterbildung
critical pigment volume concentration (CPVC): kritische Pigmentvolumenkonzentration
cross hatch adhesion test: Gitterschnitt
cross-linking: Vernetzung
cross-linking density: Vernetzungsdichte
cross-linking reaction: Vernetzungsreaktion
cross-linking velocity: Vernetzungsgeschwindigkeit
curtain coating: Gießlackieren
curtaining: Gardinenbildung
cyclised rubber: Cyclokautschuk

D

danger classes: Gefahrenklassen
danger of explosion: Explosionsgefahr
dangerous freights: Gefahrgüter
darkening: Nachdunkeln
daylight fluorescent paints: Tagesleuchtfarben
daylight fluorescent pigments: Tagesleuchtpigmente
DD-coating: DD-Lack
decomposition temperature: Zersetzungstemperatur
decorative paints: Malerlacke
defective colour vision: Farbfehlsichtigkeit
defoaming agents: Entschäumer

degreasing: Entfetten
degree of fineness: Feinheitsgrad
degree of grinding: Körnigkeit
demineralised water: entionisiertes Wasser
density: Dichte
dew point: Taupunkt
diatomaceous earth: Diatomeenerde
dibutyl phthalate (DBP): Dibutylphthalat
dicarboxylic acids: Dicarbonsäuren
DIN-chromatic-system: DIN-Farbsystem
dioctyl phthalate: Dioctylphthalat
dipole: Dipol
dipping: Tauchlackieren
dirt repellent: schmutzabweisend
di(s)azo pigments: Di(s)azopigmente
dispersing units: Dispergieraggregate
dispersion: Dispersion
dispersion paints: Dispersionsfarben
dispersion process: Dispergierung
dissolver: Dissolver
dissolving power: Lösevermögen
doctor blade: Rakel
dolomite: Dolomit
domestic appliance coatings: Hausgerätelacke
double bond: Doppelbindung
driers: Trockenstoffe
driers, dosing of: Sikkativierung
dripping resins: Träufelharze
dry film thickness: Trockenfilmdicke
drying and hardening methods: Trocknungs- und Härtungsverfahren
drying, chemical: Trocknung, chemisch
drying, forced: Trocknung, forciert
drying oils: trocknende Öle
drying, oxidativ: Trocknung, oxidativ
drying, physical: Trocknung, physikalisch
drying time measurement: Trocknungsprüfung
dry sanding: trockenschleifen
dry spraying: nebeln

dust explosion: Staubexplosion
dust-free: staubfrei
dust-free drying: Staubtrocknung
dust mask: Staubmaske
dye stuffs: Farbstoffe

E

earthing: Erdung
eco friendly: umweltfreundlich
edge covering: Kantenabdeckung
effect finishes: Effektlacke
effect pigments: Effektpigmente
elaeostearic acid: Eläostearinsäure
elastomers: Elastomere
electrical conductivity: Leitfähigkeit, elektrische
electrical insulation coatings: Elektroisolierlacke
electrically conducting varnishes: Leitlacke
electrochemical series: Spannungsreihe, elektrochemische
electrodeposition coating: Elektrotauchlackierung
electrodeposition coating, anodic: Elektrotauchlackierung, anodisch
electrodeposition coating, cathodic: Elektrotauchlackierung, kathodisch
electron beam curing (EBC): Elektronenstrahlenhärtung
electrostatic application: elektrostatische Applikation
electrostatic charge: Aufladung, elektrostatisch
embrittlement: Versprödung
emission: Emission
emulsifying agent: Emulgator
emulsion: Emulsion
environmental protection: Umweltschutz
epoxy resin esters: Epoxidharzester
epoxy resin powders: Epoxidharzpulver
epoxy resins: Epoxidharze
EPS-coating: EPS-Lackieren
Erichsen indentation: Erichsentiefung
esterification: Veresterung
esters: Ester
etherification: Veretherung

evaporate: Verdunsten
evaporation index: Verdunstungszahl
exhaust system: Absaugung
explosion limits: Explosionsgrenzen
extenders: Füllstoffe, Verschnittmittel
extruder: Extruder

F

fading: Ausbleichen
Faraday cages: Faradayscher Käfig
fatty acids: Fettsäuren
fatty acids, saturated: Fettsäuren, gesättigte
fatty acids, unsaturated: Fettsäuren, ungesättigte
fillers: Füllstoffe
filling: Abfüllung
filling primers: Grundierfüller
film: Film
film formers: Filmbildner
film forming: Verfilmung
film properties: Filmeigenschaften
film thickness measurement: Schichtdickenmessung
filtration: Filtration
fineness of grind: Mahlfeinheit
fire prevention paints: Brandschutzfarben
fire protection: Brandschutz
fire-resistant paints: flammhemmende Beschichtungen
flaking: Abblättern
flame cleaning: Flammentrostung
flammable: entzündlich
flash off: Ablüften
flash point: Flammpunkt
flatting agents: Mattierungsmittel
flexibility: Biegefestigkeit
floating: Ausschwimmen
flocculation: Flockulation
floor varnishes: Fußbodenlacke
flop: Flopeffekt
Florida test: Florida-Test

flow coating: Flow-Coating, Flutlackierung
flow cups: Auslaufbecher
flow properties: Fließeigenschaften
fluidised bed process: Wirbelsinterverfahren
fluosilicate treatment: fluatieren
foam: Schaum
forced drying: forcierte Trocknung
formaldehyde: Formaldehyd
formulation: Rezeptur
formulation development: Rezeptierung
free-flowing properties: Rieselfähigkeit
full tone colour: Volltonfarbe
furnace carbon black: Flammruß
furniture coatings: Möbellacke

G

gas-chromatography: Gas-Chromatographie
gassing: gasen
gelling power: Geliervermögen
German Standards Institute (DIN): DIN (Deutsches Institut für Normung)
GHS (Globally Harmonized System of Classification and Labelling of chemicals: Global harmonisiertes System zur Einstufung und Kennzeichnung von Chemikalien
glazes: Lasuren
gloss: Glanz
gloss measurement: Glanzmessung
gloss retention: Glanzbeständigkeit
glycerine (propanetriol): Glyzerin
glycol-ether-esters: Glykoletherester
glycol-ethers: Glykolether
gold bronze powders: Goldbronzen
gold coatings: Goldlacke
grindability: Schleifbarkeit
grind gauge: Grindometer
grinding media: Mahlkörper
grittiness: Körnigkeit
grounding: Erdung

H

hammer finish paint: Hammerschlaglack
hardener: Härter
hardness: Härte
hardness measurement: Härtemessung
harmful: gesundheitsschädlich
haze: Schleierbildung, Trübung
heavy metals: Schwermetalle
hiding power: Deckvermögen
heat stability: Hitzebeständigkeit
high boiling solvents: Hochsieder
high dispersed silicic acids: hochdisperse Kieselsäure
high pressure spraying: Hochdruckspritzen
high solids: High-Solids
hot dip galvanising: Feuerverzinkung
hot spraying: Heißspritzverfahren
house paints: Bautenlacke
hue: Farbton
HVLP: HVLP
hydrogen bridge bond: Wasserstoffbrückenbindung
hydrogen ions concentration: Wasserstoffionenkonzentration
hydrophilic: hydrophil
hydrophobic: hydrophob
hydroxyl group: Hydroxylgruppe

I

ignition temperature: Zündtemperatur
illuminant: Lichtart
immission protection: Immissionsschutz
impact test: Schlagtiefung
impregnating varnishes: Tränklacke
incineration, catalytic: Nachverbrennung, katalytische
incompatibility: Unverträglichkeit
inconbustible: nicht brennbar
indanthrene blue: Indanthrenblau
indentation hardness: Eindringhärte

industrial coatings: Industrielacke
inert pigment: inertes Pigment
infrared drying: Infrarot-Trocknung
inhibitors: Inhibitoren
inhibitory effect of oxygen: Sauerstoffinhibierung
initial drying: Antrocknung
inorganic chemistry: anorganische Chemie
inorganic pigments: anorganische Pigmente
interfacial tension: Grenzflächenspannung
interference pigments: Interferenzpigmente
interior can coatings: Doseninnenschutzlacke
interlayer adhesion: Zwischenschichthaftung
intermediate sanding: Zwischenschliff
ion exchanger: Ionenaustauscher
ionic and non-ionic: ionogen und nicht
iron blue: Eisencyanblau
iron oxide pigments: Eisenoxidpigmente
IR-radiator: IR-Strahler
irritant: reizend
isocyanates: Isocyanate
isopropanol: Isopropanol
ISO-standards: ISO-Normen

K

kaolin: Kaolin
Kesternich-test: Kesternich-Test
ketone resins: Ketonharze
ketones: Ketone
kneaders: Kneter

L

lacquer: Lack
lead chromates: Bleichromate
lead-free: bleifrei
lead molybdates: Bleimolybdate
leafing bronzes: Leafing-Bronzen

let down: Komplettierung
levelling: Verlauf
levelling agents: Verlaufsmittel
lifting: Hochziehen
light absorption: Lichtabsorption
light fastness: Lichtbeständigkeit
light refraction: Lichtbrechung
light scattering: Lichtstreuung
light stabilisers: Lichtschutzmittel
light stability: Lichtbeständigkeit
linoleic acid: Linolsäure
linolenic acid: Linolensäure
linseed oil: Leinöl
lithopone: Lithopone
long oil: langölig
lotus effect: Lotos-Effekt
low-boiling solvents: Niedrigsieder
low-pressure spraying: Niederdruckspritzen
low-solvent: lösemittelarm

M

making up: Auflacken
maleic acid: Maleinsäure
maleic resins: Maleinatharze
manual rust removal: Handentrostung
maritime climate: Seeklima
mass tone: Volltonfarbe
medium boiling solvents: Mittelsieder
medium oil: mittelölig
melamine resins: Melaminharze
melting point: Schmelzpunkt
melting range: Schmelzbereich
mesh size: Maschenweite
metallic coatings: Metallics
metallic effect pigments: Metalleffektpigmente
metal, noble: Metall, edles
metal, non-noble: Metall, unedles
metamerism: metamerie

methoxy propanol: Methoxypropanol
methyl cellulose: Methylcellulose
methylene chloride: Methylenchlorid
mica: Glimmer
micaceous iron oxide: Eisenglimmer
migration: Ausschwitzen
mill base: Mahlgut
milling: Mahlung
milling facilities: Mahlaggregate
mills: Mühlen
Milori blue: Miloriblau
minimum film-forming temperature: Mindestfilmbildungstemperatur (MFT)
minium: Bleimennige
mixing coating: Mischlack
mixing system: Mischsystem
mixture with white: Weißabmischung
moisture curing: Feuchtigkeitshärtung
moisture-induced corrosion: Feuchtigkeitskorrosion
molecule: Molekül
molybdate red: Molybdatrot
monomers: Monomere
MWC-value (maximum working conditions value): MAK-Wert

N

nail lacquers: Nagellacke
nanotechnology: Nanotechnologie
natural resins: Naturharze
needle-shaped: nadelförmig
Newtonian liquids: Newtonsche Flüssigkeiten
nickel titanium yellow: Nickeltitangelb
nitrocellulose: Nitrocellulose
nitrocellulose combination lacquers: Nitrokombinationslacke
non-aqueous dispersions: nichtwässrige Dispersionen
non-drying oils: nichttrocknende Öle
non-ferrous metals: Nichteisenmetalle
non-leafing bronzes: Non-Leafing-Bronzen
non-solvent: Nichtlöser

non-volatile matter: nichtflüchtiger Anteil
normal colour vision: Normalsichtigkeit
novolaks: Novolacke

O

occupational safety: Arbeitssicherheit
oil absorption value: Ölzahl
oil boiling: Ölverkochung
oil content: Ölgehalt
oil paints: Ölfarben
oil requirement: Ölbedarf
oils: Öle
oil varnishes: Öllacke
oleophilic: oleophil
oleophobic: oleophob
oligomers: Oligomere
one-coat systems: Einschichtsysteme
open pore: offenporig
optical properties: optische Eigenschaften
orange peel effect: Orangenhauteffekt
organic chemistry: organische Chemie
organic pigments: organische Pigmente
organosols: Organosole
outdoor stability: Außenbeständigkeit
outdoor weathering: Außenbewitterung
outturn sample: Ausfallmuster
overcoatability: Überlackierbarkeit
overcuring: Übervernetzung
overspray: Overspray
overstoving: Überbrennen
oxidation: Oxidation
oxidative drying: oxidative Trocknung

P

packing density: Packungsdichte
paint and coating drying: Lacktrocknung

paint remover: Abbeizmittel
paint system: Anstrichsystem
parameters: Parameter
parquet sealing coat: Parkettversiegelung
particle: Teilchen
particle shape: Teilchenform
particle size: Teilchengröße
particle size distribution: Korngrößenverteilung
passivating: passivieren
pearlescent pigments: Perlglanzpigmente
pencil hardness: Bleistifthärte
pendulum hardness: Pendelhärte
pentaerithritol: Pentaerythrit
peroxides: Peroxide
phase: Phase
phase interface: Phasengrenzfläche
phase separation: Phasentrennung
phenolic resins: Phenolharze
phosphate pigments: Phosphatpigmente
phosphate plasticisers: Phosphatweichmacher
phosphating: phosphatieren
phosphoric acid: Phosphorsäure
photochemical degradation: photochemischer Abbau
phthalate plasticisers: Phthalatweichmacher
phthalic acid: Phthalsäure
phthalocyanine blue pigments: Phthalocyaninblau-Pigmente
phthalocyanine green pigments: Phthalocyaningrün-Pigmente
pH-value: pH-Wert
physical drying: Trocknung, physikalisch
physical properties: physikalische Eigenschaften
physiological harmlessness: physiologische Unbedenklichkeit
pigmentation: Pigmentierung
pigment content: Pigmentgehalt
pigment lakes: Verlacken von Farbstoffen
pigment paste: Pigmentpaste
pigment preparation: Pigmentpräparation
pigment properties: Pigmenteigenschaften
pigments: Pigmente
pigment surface: Pigmentoberfläche

pigment testing: Pigmentprüfung
pigment volume concentration (PVC): Pigmentvolumenkonzentration
pigment wetting: Pigmentbenetzung
pilot plant: Technikum
pinholes: Nadelstiche
plain colour coating: Uni-Lackierung
plasma treatment: Plasmabehandlung
plasticisers: Weichmacher
plastics, coatings for: Kunststofflacke
plastification: Plastifizieren
plastification, external: Weichmachung, äußere
plastification, internal: Weichmachung, innere
plastisols: Plastisole
plate-like: schuppenförmig
polarity: Polarität
polishing: polieren
polyacrylic resins: Polyacrylatharze
polyaddition: Polyaddition
polyalcohols: Polyalkohole
polyamides: Polyamide
polyamines: Polyamine
polycarboxylic acids: Polycarbonsäuren
polycondensation: Polykondensation
polyester putties: Polyesterspachtel
polyester resins: Polyesterharze
polyester resins, unsaturated: Polyester, ungesättigt
polyisocyanates: Polyisocyanate
polyisocyanates, blocked: Polyisocyanate, blockiert
polymer: Polymer
polymer dispersion: Kunststoff-Dispersion
polymerisation: Polymerisation
polysiloxanes: Polysiloxane
polystyrene: Polystyrol
polyurethanes (PUR): Polyurethan (PUR)
polyvinyl acetate: Polyvinylacetat (PVA)
polyvinyl alcohol: Polyvinylalkohol (PVAL)
polyvinyl butyral (PVB): Polyvinylbutyral (PVB)
polyvinyl chloride (PVC): Polyvinylchlorid (PVC)
polyvinyl propionate: Polyvinylpropionat

porosity: Porosität
porosity test: Porositätsmessung
pot life: Topfzeit
powder coating process: Pulverbeschichten
powder coating production: Pulverlackherstellung
powder coatings: Pulverlacke
powder spraying, electrostatic: Pulversprühen, elektrostatisch
pre-dispersion: Vordispergieren
pre-mixing: Ansetzerei
preservative impregnation for wood: Holzschutzimprägnierung
pre-treatment methods: Vorbehandlungsverfahren
pre-treatment of metals: Metallvorbehandlung
pre-treatment of substrates: Untergrundvorbehandlung
pre-treatment of surfaces: Oberflächenvorbehandlung
prevention of air pollution: Luftreinhaltung, Immissionsschutz
primary particles: Primärteilchen
primers: Grundierungen
primer surfacers: Füller
production batch: Ansatz
propanol: Propanol
propellant gas: Treibgas
protective equipment: Schutzausrüstung
protective gloves: Schutzhandschuhe
protective goggles: Schutzbrille
protective shoes: Sicherheitsschuhe
pseudoplasticity: Strukturviskosität
putty: Spachtel
PVC: Pigmentvolumenkonzentration
pycnometer: Pyknometer

Q

quality control: Qualitätskontrolle
quality management: Qualitätsmanagement
quinacridones: Chinacridone
q-value: Q-Wert

R

radiation: Strahlung
radiation curing: Strahlenhärtung
radiator enamels: Heizkörperlacke
radical: Radikal
RAL-register: RAL-Register
rapid-rotation atomisers: Hochrotationszerstäuber
raw materials for paints and coatings: Lackrohstoffe
REACH (Registration, Evaluation, Authorisation of Chemicals): Registrierung, Evaluierung, Authorisierung von Chemikalien
reaction: Reaktion
reactive pigment: actives Pigment
reactive primer: Reaktionsgrund
reactive thinners: Reaktivverdünner
recycling: Recycling
reduction: Reduktion, Auflacken
reflection curve: Reflexionskurve
reflectometer: Reflektometer
refractive index: Brechzahl
refractometer: Refraktometer
regeneration: regenerieren
relative humidity: relative Luftfeuchtigkeit
release agents: Trennmittel
reproducibility: Reproduzierbarkeit
resins: Harze
resistance: Beständigkeit
resistance, electrical: Widerstand, elektrischer
resite: Resit
resole: Resol
retention: Retention
revolutions per minute: Umdrehungszahl
rheological additives: Rheologie-Additive
rheology: Rheologie
ricinolic acid: Ricinolsäure
ring circuit stability: Ringleitungsstabilität
road line paints: Straßenmarkierungsfarben
rod bending test: Dornbiegeprüfung
roller: Rolle

roller coating: Walzlackierung
room temperature: Raumtemperatur
rotation viscometers: Rotationsviskosimeter
roughness height: Rautiefe
roughness measurement: Rauheitsbestimmung
rubber derivatives: Kautschukderivate
rub out test: Rub-out-Test
rust: Rost
rust prevention primers: Rostschutzgrundierung
rust protection: Rostschutz
rust removal: Rostentfernung
rutile: Rutil

S

safety: Arbeitssicherheit
safety data sheet: Sicherheitsdatenblatt
sag: Gardinenbildung
sagging disposition: Ablaufneigung
sagging resistance: Standfestigkeit
salt efflorescence: Salzausblühung
salt spray test: Salzsprühtest
sand blasting: Sandstrahlen
sanding: schleifen
sanding dust: Schleifstaub
sanding machines, manual: Rutscher
sanding paper: Schleifpapier
sanding properties: Schleifbarkeit
sand mill: Sandmühle
saponifiability: Verseifbarkeit
saponification value: Verseifungszahl
saturated atmosphere test: Schwitzwassertest
scale: Zunder
scattering power: Streuvermögen
scratch resistance: Kratzfestigkeit
scratch test: Ritzhärte
screen analysis: Siebanalyse
screening: Filtration
scrub resistance: Abriebbeständigkeit

sealer: Sperrgrund
sediment: Bodensatz
settling: Absetzen
shear rate: Schergeschwindigkeit
shellac: Schellack
shop primers: Shop-Primers
short oil: kurzölig
shrinkage: Schrumpfung
sieve: Sieb
sieve analysis: Siebanalyse
silica flour: Quarzmehl
silicic acid esters: Kieselsäureester
silicic acid, precipitated: Kieselsäure, gefällte
silicic acid, pyrogenic: Kieselsäure, hochdisperse
silicification: Verkieselung
silicone dioxide: Siliciumdioxid
silicone oils: Siliconöle
silicone polyester resins: Siliconpolyester
silicone resins: Siliconharze
silicones: Silicone
skinning: Hautbildung
skull and crossbones: Totenkopf
slip additives: Slip-Additive
slipperiness: Gleitfähigkeit
slurry process: Slurry-Verfahren
smear point: Schmierpunkt
soft resins: Weichharze
solids content (non-volatile matter): Festkörper
solubility: Löslichkeit
solution, colloidal: Lösung, kolloidale
solution, true: Lösung, echte
solvation: Solvatation
solvation wrapping: Solvatationshülle
solvent-free: lösemittelfrei
solvent, latent: Löser, latenter
solvent, real: Löser, echter
solvent resistance: Lösemittelbeständigkeit
solvent retention: Lösemittelretention
solvents: Lösemittel

solvent shock: Lösemittelschock
soya bean oil: Sojaöl
special waste incineration: Sondermüllverbrennung
specifications: Spezifikationen
specific surface air: spezifische Oberfläche
speck formation: Stippenbildung
spectrophotometer: Spektralphotometer
spectrum colours: Spektralfarben
spinel pigments: Spinellpigmente
spot putty: Fleckspachtel
spot repair: Beispritzen
spray application: Spritzapplikation
spray application processes: Spritzverfahren
spray booth: Spritzkabine
spray dust: Spritznebel
spray guns: Spitzpistolen
spraying automatic devices: Spritzautomaten
spraying pattern: Spritzbild
spraying station: Spritzstand
spraying thinner: Spritzverdünnung
spreading: Spreitung
spreading rate: Ergiebigkeit
stabilisation (of dispersions): Stabilisierung (von Dispersionen)
staining: beizen
stainless steel: rostfreier Stahl
standardisation: Standardisierung
standard sample: Standardmuster
stand oil: Standöl
stearic acid: Stearinsäure
steel: Stahl
sticky tape detachment: Klebeband-Abreißmethode
stirrers: Rührer
stone chipping protection: Steinschlagschutz
stone chipping test: Steinschlagtest
storage stability: Lagerstabilität
stoving: Ofentrocknung
stoving coatings: Einbrennlacke
stoving temperature: Einbrenntemperatur
stress: Schubspannung

structured paints: Strukturlacke
styrenated alkyd resins: Styrolalkydharze
styrene: Styrol
subsequent cross-linking: Nachvernetzung
substrate: Untergrund
sub-surface rusting: Unterrostung
subtraktive colour mixture: subtraktive Farbmischung
suggested formulations: Richtrezepturen
sun flower oil: Sonnenblumenöl
surface active: oberflächenaktiv
surface area, specific: Oberfläche, spezifische
surface cleaning: Oberflächenreinigung
surface colour: Körperfarbe
surface damages: Oberflächenstörungen
surface protection: Oberflächenschutz
surfacers: Füller
surface tension: Oberflächenspannung
surfactants: Tenside
suspension: Suspension
swelling: Anquellen
synthesis: Synthese
synthetic: synthetisch
synthetic resins: Kunstharze

T

talc: Talkum
talloil: Tallöl
tank farm: Tanklager
temperature: Temperatur
terephthalic acid: Terephthalsäure
terpenoid hydrocarbons: Terpenkohlenwasserstoffe
test methods: Prüfverfahren
test specification: Prüfvorschriften
thermoplastic resins: Thermoplaste
thermosets: Duroplaste
thermosetting: Ofentrocknung
thickener: Verdicker
thickening: Eindicken

thinnability: Verdünnbarkeit
thinner: Verdünnungsmittel
thixotropic agents: Thixotropierungsmittel
thixotropy: Thixotropie
through curing: Durchhärtung
throwing power: Umgriff
tinting: abtönen, nuancieren
tinting base: Abtönpaste
tinting strength: Farbstärke, Färbevermögen
titanium dioxide: Titandioxid
titration: Titration
tolerance: Toleranz
toluene: Toluol
toluolsulphonic acid: Tolouolsulphonsäure
top coat: Decklack
toxicity: Toxizität
transparency: Transparenz
treatment of sewage: Abwasserreinigung
tribo-application: Triboapplikation
trilene dipping process: Tri-Tauchverfahren
trimellitic acid: Trimellithsäure
trimethylol propane: Trimethylolpropan
triple roll mill: Dreiwalze
tristimulus method: Tristimulusverfahren
tristimulus values: Normfarbwerte
trivalent alcohols: dreiwertige Alkohole
trowelling: Ziehen (von Spachteln)
tube enamels: Tubenemaillen
turpentine oil: Terpentinöl
turpentine oil substitute: Terpentinersatz
two-coat metallic coatings: Metallic-Zweischichtlackierung
two-coat systems: Zweischichtlackierung
two-component systems: Zweikomponenten-Systeme
two-pack systems: Zweikomponenten-Systeme

U

ultra-filtration: Ultrafiltration
ultramarine pigments: Ultramarin-Pigmente

ultraviolet radiation: ultraviolette Strahlung
underbody sealing: Unterbodenschutz
undercoat varnish: Vorlack
undercuring: Untervernetzung
underwater coatings: Unterwasseranstriche
unsaturated fatty acids: ungesättigte Fettsäuren
unsaturated polyesters: ungesättigte Polyesters
urea resins: Harnstoffharze
urethane: Urethan
UV-absorbers: UV-Absorber
UV-curing: UV-Härtung

V

vacuum dissolver: Vakuum-Dissolver
vacuum impregnation: Vakuum-Imprägnierung
value: Kennzahl
Van der Waals forces: van-der-waalsche Kräfte
vapour cure process: Vapour Cure-Verfahren
ventilation: Belüftung
ventilation surfacer: Ventilationsgrund
vibrating screen: Schüttelsieb
vinyl acetate: Vinylacetat
vinyl chloride: Vinylchlorid
vinyl propionate: Vinylpropionat
viscometers: Viskosimeter
viscosity: Viskosität
viscosity anomalies: Viskositätsanomalien
visible light: sichtbares Licht
visual evaluation: visuelle Beurteilung
VOC-directive: VOC-Richtlinie
volatile matter: flüchtiger Anteil
Vossen blue: Vossen Blau

W

wall paints: Wandfarben
wall paints, exterior: Fassadenfarben

wall paints, interior: Innenwandfarben
wash primers: Wash-Primer
waste disposal: Abfallbeseitigung
water-borne coatings: Wasserlacke
water, desalinated: entionisiertes Wasser
water-free: wasserfrei
water glass paints: Wasserglasanstrichfarben
water hardness: Wasserhärte
water mountain: Wasserberg
water repellent (hydrophobic): wasserabweisend
water resistance: Wasserbeständigkeit
water solubility: Wasserlöslichkeit
water thinnability: Wasserverdünnbarkeit
water treatment: Wasseraufbereitung
water vapour permeability: Wasserdampfdurchlässigkeit
wavelength: Wellenlänge
waxes: Wachse
weathering, accelerated: Kurzbewitterung
weathering, artificial: Bewitterung, künstliche
weathering station: Bewitterungsstation
weather-o-meter: Weather-O-Meter (WOM)
weather resistance: Wetterbeständigkeit
wet abrasion resistance: Scheuerfestigkeit
wet film thickness: Nassfilmdicke
wet-on-wet-application: Nass-in-Nass-Verfahren
wet rinsing systems: Wasserberieselung
wet-rub resistance: Nasswischfestigkeit
wet sanding: nassschleifen
wet sieving: Nass-Siebung
wetting: benetzen
wetting agents: Netzmittel
white goods: weiße Ware
whitening power: Aufhellvermögen
white pigments: Weißpigmente
white rust: Weißrost
white spirits: Testbenzin
white standard: Weißstandard
wood coatings: Holzlacke
wood glazes: Holzlasuren

wood moisture: Holzfeuchtigkeit
wood oil: Holzöl
wood rosin: Wurzelharz
wood stain: Holzbeize
wool scale: Wollskala
wrinkle finishes: Runzellacke
wrinkling: Runzelbildung

X

xenon test: Xenontest
xylene: Xylol
xylene/formaldeyde resins: Xylol/Formaldehyd-Harze

Y

yellowing: Vergilbung
yield: Ergiebigkeit
yield point: Fließpunkt

Z

zapon lacquers: Zaponlacke
zeolites: Zeolithe
zeta potential: Zetapotential
zinc (to): verzinken
zinc chromate: Zinkchromat
zinc dust: Zinkstaub
zinc dust primers: Zinkstaubgrundierungen
zinc oxide: Zinkoxid
zinc phosphate: Zinkphosphat
zinc phosphating: Zinkphosphatierung
zinc sulphide: Zinksulfid
zinc tetraoxichromate: Zinktetraoxichromat
zinc white: Zinkweiß
zirconium drier: Zirkoniumtrockenstoff

A

Abbeizmittel: paint remover
Abblättern: flaking
Abfallbeseitigung: waste disposal
Abfüllung: filling
Ablaufneigung: sagging disposition
Ablüften: flash off
Absaugung: exhaust system
Absetzen: settling
Abriebbeständigkeit: abrasion resistance
abtönen: tinting
Abtönpaste: tinting base
Abwasserreinigung: treatment of sewage
Acrylatharze: acrylic resins
Additionsreaktion: addition reaction
Additiv: additive
Additive Farbmischung: additive colour mixture
Adhäsion: adhesion
Aerosol: aerosol
Agglomerate: agglomerates
Aggregate: aggregates
Airbrush-Technik: airbrush-technique
Airless-Spritzen: airless spraying
Airmix-Spritzen: airmix spraying
aktives Pigment: reactive pigment
aliphatische Kohlenwasserstoffe: aliphatic hydrocarbons
Alkalibeständigkeit: alkali resistance
Alkohole: alcohols
Alkyharze: alkyd resins
Alkydharze, kurzölig: alkyd resins, short oil
Alkydharze, langölig: alkyd resins, long oil
Alkydharze, mittelölig: alkyd resins, medium oil
Aluminium: aluminium
Aluminiumbronze: aluminium bronze
Amine: amines
Aminoharze: amino resins
Anatas: anatase
Anion: anion

Anode: anode
anorganische Chemie: inorganic chemistry
anorganische Pigmente: inorganic pigments
anquellen: swelling
Ansatz: production batch
Ansetzerei: pre-mixing and pre-dispersion
Anstrichsystem: paint system
Antiabsetzmittel: anti-settling agents
Antiausschwimmmittel: anti-floating agents
Antistatika: anti-static agents
Antrocknung: initial drying
anwendungstechnische Eigenschaften: application technological properties
Applikationsverfahren: application methods
Aräometer: areometer
Arbeitssicherheit: occupational safety
Aromaten: aromatic hydrocarbons
aromatenfrei: aromatic-free
Atemschutz: breathing protection
Aufhellvermögen: whitening power
Auflacken: reduction, making up
Aufladung, elektrostatisch: electrostatic charge
Ausbleichen: fading
Ausbluten: bleeding
Ausfallmuster: outturn sample
Aushärtung: curing
Auslaufbecher: flow cups
Ausschwimmen: floating
Ausschwitzen: migration
Außenbeständigkeit: outdoor resistance
Außenbewitterung: outdoor weathering
Autolackierung: automotive coating
Autoreparaturlacke: automotive refinishing coatings
Autoreparaturlackierung: automotive refinishing
Autoreparatursysteme: automotive refinishing systems
Autoseriendecklacke: automotive (OEM) topcoats.
Autoserienfüller: automotive(OEM) fillers
Autoseriengrundierungen: automotive (OEM) primers
Azopigment: azo pigment

B

Bariumsulfat: barium sulphate
Barrierewirkung: barrier effect
Basen: bases
Bautenlacke: house paints, architectural paints
Beispritzen: spot repair
Beizen: staining
Belüftung: ventilation
Benardsche Zellen: Benard's cells
benetzen: wetting
Bentonit: bentonite
Bentonit-Paste: bentonite paste
Benzin: white spirits
Berliner Blau: Berlin blue
Beschichtungsstoffe: coating materials
Beschichtungssysteme: coating systems
Beschichtungsverfahren: application methods
Beschleuniger: accelerators
Beständigkeit: resistance
Bewitterung, künstliche: weathering, artificial
Bewitterungsstation: weathering station
Biegefestigkeit: flexibility
Bindemittel: binders
Bindemittelbedarf: binder requirement
Biozide: biocides
Bläueschutz: blue stain inhibition
Blanc-Fixe: blanc fixe
Blasenbildung: blistering
Blau-Gelb-Wert: blue-yellow-value
Bleichromate: lead chromates
bleifrei: lead-free
Bleimennige: minium
Bleimolybdate: lead molybdates
Bleistifthärte: pencil hardness
Blockfestigkeit: blocking resistance
blockiertes Isocyanat: blocked isocyanate
Bodensatz: sediment
Brandschutz: fire protection

Brandschutzfarben: fire prevention paints
Brechzahl: refractive index
Brennbarkeit: inflammability
Broncieren: bronzing
Buntpigmente: coloured pigments
Butanol: butanol
Butylacetat: butyl acetate
Butylglykol: butyl glycol

C

cadmiumfrei: cadmium-free
Cadmiumpigmente: cadmium pigments
Calcit: calcite
Calciumcarbonat: calcium carbonate
Carboxylgruppe: carboxyl group
Celluloseacetobutyrat: cellulose acetate butyrate
Cellulosenitrat: cellulose nitrate
Chemikalienbeständigkeit: chemical resistance
China Clay: china clay
Chinacridone: quinacridones
Chlorkautchuk: chlorinated rubber
Chlorparaffine: chlorinated paraffins
chromatfrei: chromate-free
chromatieren: chromating
Chromoxidgrün: chromium oxide green
Chromtitangelb: chromium titanium yellow
CIE-L*a*b-System: CIE-L*a*b system
Citronensäureester: citric acid esters
Coil-Coating: coil coating
Colöser: co-solvents
Colour-Index (CI): colour index (CI)
Copolymerisation: copolymerisation
Cyclokautschuk: cyclised rubber

D

DD-Lack: DD-coating

Decklack: top coat
Deckvermögen: hiding power
Diatomeenerde: diatomaceous earth
Dibutylphthalat: dibutyl phthalate
Dicarbonsäuren: dicarboxylic acids
Dichte: density
DIN (Deutsches Institut für Normung): German Standards Institute
DIN-Farbsystem: DIN-chromatic-system
Dioctylphthalat: dioctyl phthalate
Dipol: dipole
Di(s)azopigmente: di(s)azo pigments
Dispergieraggregate: dispersing units
Dispergierung: dispersion process
Dispersion: dispersion
Dispersionsfarben: dispersion paints
Dissolver: dissolver, agitator
Dolomit: dolomite
Doppelbindung: double bond
Dornbiegeprüfung: rod bending test
Doseninnenschutzlacke: interior can coatings
Dreiwalze: triple roll mill
dreiwertige Alkohole: trivalent alcohols
Druckluftspritzen: compressed-air spraying
Durchhärtung: through curing
Duroplaste: thermosets

E

Echtheit: fastness
Effektlacke: effect finishes
Effektpigmente: effect pigments
Einbrennen: stoving, thermosetting
Einbrennlacke: stoving coatings
Einbrenntemperatur: stoving temperature
Eindicken: thickening
Eindringhärte: indentation hardness
Einfallswinkel: angle of incidence
Eingangskontrolle: acceptance test
Einschichtsysteme: one-coat systems

Eisencyanblau: iron blue
Eisenglimmer: micaceous iron oxide
Eisenoxidpigmente: iron oxide pigments
Eläostearinsäure: elaeostearic acid
Elastomere: elastomers
Elektroisolierlacke: electrical insulation coatings
Elektronenstrahlenhärtung: electron beam curing
elektrostatische Applikation: electrostatic application
Elektrotauchlackierung (ETL): electrodeposition coating
Elektrotauchlackierung, anodisch: electrodeposition coating, anodic
Elektrotauchlackierung, kathodisch: electrodeposition coating, cathodic
Emission: emission
Emulgator: emulsifying agent
Emulsion: emulsion
entfetten: degreasing
entionisiertes Wasser: demineralised water
entrosten: rust removal
Entschäumer: defoaming agents
entzündlich: flammable
Epoxidharze: epoxy resins
Epoxidharzester: epoxy resin esters
Epoxidharzpulver: epoxy resin powders
EPS-Lackieren: EPS-coating
Erdung: earthing, grounding
Ergiebigkeit: spreading rate, yield
Erichsentiefung: Erichsen indentation
Ester: esters
Explosionsgefahr: danger of explosion
Explosionsgrenzen: explosion limits
Extruder: extruder

F

Faradayscher Käfig: Faraday cage
Farbabstand: colour difference
Farbe: colour
Farbe, bunte: colour, chromatic
Farbe, unbunte: colour, achromatic

Farbfehlsichtigkeit: defective colour vision
Farbkontrast: colour contrast
Farbkreis: colour circle
Farbmessgerät: colour measurement instrument
Farbmetrik: colur measurement
Farbmittel: colourant
Farbort: colour location
Farbrezepturberechnung: colour formulation calculation
Farbstärke: tinting strength
Farbstandard: colour standard
Farbstoffe: dyestuffs
Farbton: hue
Farbtonbeständigkeit: colour fastness
Farbzahl: colour value
Fassadenfarben: wall paints, exterior
Feinheitsgrad: degree of fineness
Festkörper: solids content (non-volatile matter)
Fettsäuren: fatty acids
Fettsäuren, gesättigt: fatty acids, saturated
Fettsäuren, ungesättigt: fatty acids, unsaturated
Feuchtigkeit, relative: relative humidity
Feuchtigkeitshärtung: moisture curing
Feuchtigkeitskorrosion: moisture induced corrosion
Feuerverzinkung: hot dip galvanising
Film: film
Filmbildner: film formers
Filmeigenschaften: film properties
Filtration: filtration
Firnis: boiled oil
Flämmen von Kunststoff: blazing of plastic surfaces
Flammentrostung: flame cleaning
flammhemmende Beschichtungen: fire-resistant paints
Flammpunkt: flash point
Flammruß: furnace carbon black
Fleckspachtel: spot putty
Fließeigenschaften: flow properties
Fließpunkt: yield point
Flockulation: flocculation
Flopeffekt: flop

Florida-Test: Florida test
Flow-Coating: flow-coating
fluatieren: fluosilicate treatment
flüchtiger Anteil: volatile matter
Flutlackierung: flow-coating
forcierte Trocknung: forced drying
Formaldehyd: formaldehyde
fotochemischer Abbau: photochemical degradation
Freibewitterung: outdoor weathering
Füller: filler, primer surfacer
Fußbodenlacke: floor varnishes

G

Gardinenbildung: curtaining
Gaschromatographie: gas-chromatography
gasen: gassing
Gasruß: channel carbon black
geblasenes Öl: blown oil
Gefahrenklassen: danger classes
Gefahrgüter: dangerous freights
Geliervermögen: gelling power
geschlossenporig: closed pore
gesundheitsschädlich: harmful
GHS (Global harmonisiertes System zur Einstufung und Kennzeichnung von Chemikalien): Globally Harmonized System of Classification and Labelling of chemicals
Gießlackieren: curtain coating
giftig: toxic
Gitterschnitt: cross hatch adhesion test
Glanz: gloss
Glanzbeständigkeit: gloss retention
Glanzmessung: gloss measurement
Gleitfähigkeit: slipperiness
Glimmer: mica
Glykolether: glycol-ethers
Glykoletherester: glycol-ether-esters
Glyzerin: glycerin
Goldbronzen: gold bronze powders

Goldlacke: gold coatings
Grenzflächespannung: interfacial tension
Grindometer: grind gauge
Grundierfüller: filling primers
Grundierungen: primers

H

Härte: hardness
Härtemessung: hardness measurement
Härter: hardener
Härtung: hardening
Haftgrund: wash-primer
Haftprimer (für Kunststoffe): adhesion primer
Haftung: adhesion, adhesive strength
Haftungsprüfung: adhesion tests
Hammerschlaglack: hammer finish paint
Handentrostung: manual rust removal
Harnstoffharze: urea resins
Harze: resins
Hausgerätelacke: domestic appliance coatings
Hautbildung: skinning
Hautverhinderungsmittel: anti-skinning agents
Heißspritzverfahren: hot spraying
Heizkörperlacke: radiator enamels
High Solids: high solids
Hilfslösemittel: auxiliary solvents
Hilfsmittel: auxiliary products
Hitzebeständigkeit: heat stability
hochdisperse Kieselsäuren: high dispersed silicic acids
Hochdruckspritzen: high pressure spraying
Hochrotationszerstäuber: rapid-rotation atomisers
Hochsieder: high boiling solvents
Hochziehen: lifting
Holzbeize: wood stain
Holzfeuchtigkeit: wood moisture
Holzlacke: wood coatings
Holzlasuren: wood glazes
Holzöl: wood oil

Holzschutzimprägnierung: preservative impregnation for wood
HVLP: HVLP
hydrophil: hydrophilic
hydrophob: hydrophobic, water repellent
Hydroxylgruppe: hydroxyl group

I

Immissionsschutz: immission protection
Indanthrenblau: indanthrene blue
Industrielacke: industrial coatings
inertes Pigment: inert pigment
Infrarot-Trocknung: infrared drying
Inhibitoren: inhibitors
Innenwandfarben: wall paints, interior
Interferenzpigmente: interference pigments
Ionenaustauscher: ion exchanger
ionogen und nicht: ionic and non-ionic
IR-Strahler: IR-radiator
Isocyanate: isocyanates
ISO-Normen: ISO-standards
Isopropanol: isopropanol

K

Kältebeständigkeit: cold resistance
Kälteprüfung (von Holzlacken): cold check test
Kantenabdeckung: edge covering
Kaolin: kaolin, china clay
Katalysatoren: catalysts
Kathode: cathode
Kation: cation
Kautschukderivate: rubber derivatives
Kennzahl: value
Kesternich-Test: Kesternich-test
Ketone: ketones
Ketonharze: ketone resins

Kieselsäure, gefällte: silicic acid, precipitated
Kieselsäure, hochdisperse: silicic acid, pyrogenic
Kieselsäureester: silicic acid esters
Klarlacke: clear coats
Klebeband-Abreißmethode: sticky tape detachment
Klimaraum: conditioned room
Kneter: kneaders
Koagulation einer Emulsion: coagulation of an emulsion
Koaleszenzhilfsmittel: coalescence agents
Kobaltblau: cobalt blue
Kobalttrockenstoff: cobalt drier
Kocher: blistering
Körnigkeit: grittiness, degree of grinding
Körperfarbe: surface colour
Kohäsion: cohesion
Kokosöl: coconut-oil
Kollodiumwolle: collodion cotton
kolloidale Lösung: colloidal solution
Kolophonium: colophony
Komplementärfarben: complementary colours
komplettieren: let down
Kondensationsharze: condensation resins
Konservendosenlacke: can coatings
Kontaktgrund: catalyst primer
Konvektionsofen: convection drier
Kornfeinheit: degree of fineness
Korngrößenverteilung: particle size distribution
Korona-Aufladung: corona charge
Korona-Entladung: corona discharge
Korrosion: corrosion
Korrosionsschutz, leicht: corrosion protection, OEM, refinish, industrial etc.
Korrosionsschutz, schwer: corrosion protection, heavy duty
Korrosionschutzgrundierungen: rust prevention primers
Korrosionsschutzpigmente: anti-corrosive pigments
Korrosionsschutzprüfungen: corrosion tests
Kraterbildung: cratering
Kratzfestigkeit: scratch resistance
Kreide: chalk

Kreidung: chalking
Kritische Pigmentvolumenkonzentration: critical pigment volume concetration
Kugelmühle: ball mill
Kunstharze: synthetic resins
Kunststoffdispersion: polymer dispersion
Kunststofflacke: coatings for plastics
Kunststofflackierung: coating of plastics
Kupferphthalocyanine: copper phthalocyanines
Kurzbewitterung: weathering, accelerated
kurzölig: short oil

L

Lack: lacquer
Lackherstellung: coating production
Lackrezeptur: coating formulation
Lackrohstoffe: raw materials for paints and coatings
Lacktrocknung: paint and coating drying
Läufer: sag
Lagerstabilität: storage stability
langölig: long oil
Lasuren: glazes
Leafing-Bronzen: leafing bronzes
Leinöl: linseed oil
Leitfähigkeit, elektrische: electrical conductivity
Leitlacke: electrically conducting varnishes
Lichtabsorption: light absorption
Lichtart: illuminant
Lichtbeständigkeit: light fastness, light stability
Lichtbrechung: light refraction
Lichtdurchlässigkeit: transparency
Lichtschutzmittel: light stabilisers
Lichtstreuung: light scattering
Linolensäure: linolenic acid
Linolsäure: linoleic acid
Lithopone: lithopone
Lösemittel: solvents
lösemittelarm: low-solvent

Lösemittelbeständigkeit: solvent resistance
lösemittelfrei: solvent-free
Lösemittelretention: solvent retention
Lösemittelschock: solvent shock
Löser, echter: solvent, real
Löser, latenter: solvent, latent
Lösevermögen: dissolving power
Löslichkeit: solubility
Lösung, echte: solution, true
Lösung, kolloidale: solution, colloidal
Lotos-Effekt: lotus effect
Luftreinhaltung: prevention of air pollution

M

Mahlaggregate: milling facilities
Mahlfeinheit: fineness of grind
Mahlgut: mill base
Mahlkörper: grinding media
Mahlung: milling
MAK-Wert: MWC-value
Maleinatharze: maleic resins
Maleinsäure: maleic acid
Malerlacke: decorative paints
Maschenweite: mesh size
Mattierungsmittel: flatting agents
Melaminharze: melamine resins
Metallbandlackierung: coil coating
Metall, edles: metal, noble
Metalleffektpigmente: metallic effect pigments
Metallics: metallic coatings
Metallic-Zweischichtlackierung: two-coat metallic coatings
Metall, unedles: metal, non-noble
Metallvorbehandlung: pre-treatment of metals
Metamerie: metamerism
Methoxypropanol: methoxy propanol
Methylcellulose: methyl cellulose
Methylenchlorid: methylene chloride
Miloriblau: Milori blue

Mindestfilmbildungstemperatur: minimum filmforming temperature
Mischlack: mixing coating
Mischsystem: mixing system
mittelölig: medium oil
Mittelsieder: medium boiling solvents
Möbellacke: furniture coatings
Molekül: molecule
Molybdatrot: molybdate red
Monomere: monomers
Mühlen: mills

N

Nachbehandlung: after-treatment
Nachdunkeln: darkening
Nachverbrennung, katalytische: incineration, catalytic
Nachvernetzung: subsequent cross-linking
nadelförmig: needle-shaped
Nadelstiche: pinholes
Nagellacke: nail lacquers
Nanotechnologie: nanotechnology
Nassfilmdicke: wet film thickness
Nass-in-Nass-Verfahren: wet-on-wet-application
nassschleifen: wet sanding
Nass-Siebung: wet sieving
Nasswischfestigkeit: wet-rub resistance
Naturharze: natural resins
nebeln: dry spraying
Netzmittel: wetting agents
Newtonsche Flüssigkeiten: Newtonian liquids
nicht brennbar: incombustible
Nichteisenmetalle: non-ferrous metals
Nichtlöser: non-solvent
nichttrocknende Öle: non-drying oils
nichtwässrige Dispersionen: non-aqueous dispersions
Nickeltitangelb: nickel titanium yellow
Niederdruckspritzen: low-pressure spraying
Niedrigsieder: low-boiling solvents

Nitrocellulose: nitrocellulose
Nitrokombinationslacke: nitrocellulose combination lacquers
Non-Leafing-Bronzen: non-leafing bronzes
Normalsichtigkeit: normal colour vision
Normfarbtafel: chromaticity diagram
Normfarbwerte: tristimulus values
Novolacke: novolaks
nuancieren: tinting

O

Oberfläche, spezifische: surface area, specific
oberflächenaktiv: surface active
Oberflächenreinigung: surface cleaning
Oberflächenschutz: surface protection
Oberflächenspannung: surface tension
Oberflächenstörungen: surface damages
Oberflächenvorbehandlung: pre-treatment of surfaces
Ölbedarf: oil requirement
Öle: oils
Ölfarben: oil paints
Ölfirnis: boiled oil
Ölgehalt: oil content
Öllacke: oil varnishes
Ölverkochung: oil boiling
Ölzahl: oil absorption value
Ofentrocknung: stoving, thermosetting
offenporig: open pore
oleophil: oleophilic
oleophob: oleophobic
Oligomer: oligomer
optische Eigenschaften: optical properties
Orangenhauteffekt: orange peel effect
organische Chemie: organic chemistry
organische Pigmente: organic pigments
Organosole: organosols
Overspray: overspray
Oxidation: oxidation
oxidative Trocknung: oxidative drying

P

Packungsdichte: packing density
Parameter: parameters
Parkettversiegelung: parquet sealing coat
passivieren: passivating
Pendelhärte: pendulum hardness
Pentaerythrit: pentaerythritol
Perlglanzpigmente: pearlescent pigments
Perlmühlen: bead mills
Peroxide: peroxides
Phase: phase
Phasengrenzfläche: phase interface
Phasentrennung: phase separation
Phenolharze: phenolic resins
phosphatieren: phosphating
Phosphatpigmente: phosphate pigments
Phosphatweichmacher: phosphate plasticisers
Phosphorsäure: phosphoric acid
photochemicher Abbau: photochemical degradation
Phthalatweichmacher: phthalate plasticisers
Phthalocyaninblaupigmente: phthalocyanine blue pigments
Phthalocyaningrünpigmente: phthalocaynine green pigments
Phthalsäure: phthalic acid
pH-Wert: pH-value
physikalische Eigenschaften: physical properties
physikalische Trocknung: physical drying
physiologische Unbedenklichkeit: physiological harmlessness
Pigmente: pigments
Pigmentbenetzung: pigment wetting
Pigmenteigenschaften: pigment properties
Pigmentgehalt: pigment content
Pigmentierung: pigmentation
Pigmentöberfläche: pigment surface
Pigmentpaste: pigment paste
Pigmentpräparation: pigment preparation
Pigmentprüfung: pigment testing
Pigmentvolumenkonzentration: pigment volume concentration (PVC)
Pinsel: brush

PKW-Lackierung: automotive coating
Plasmabehandlung: plasma treatment
plastifizieren: plastification
Plastisole: plastisols
Polarität: polarity
polieren: polishing
Polyacrylatharze: polyacrylic resins
Polyaddition: polyaddition
Polyalkohole: polyalcohols
Polyamide: polyamides
Polyamine: polyamines
Polycarbonsäuren: polycarboxylic acids
Polyester, ungesättigt: polyester resins, unsaturated
Polyesterharze: polyester resins
Polyesterspachtel: polyester putties
Polyisocyanates: polyisocyanates
Polyisocyanate, blockiert: polyisocyanates, blocked
Polykondensation: polycondensation
Polymer: polymer
Polymerisation: polymerisation
Polysiloxane: polysiloxanes
Polystyrol: polystyrene
Polyurehane: polyurethanes (PU)
Polyvinylacetat: polyvinyl acetate
Polyvinylalkohol: polyvinyl alcohol
Polyvinylbutyral: polyvinyl butyral (PVB)
Polyvinylchlorid: polyvinyl chloride (PVC)
Polyvinylpropionat: polyvinyl propionat
Porosität: porosity
Porositätsmessung: porosity test
Primärteilchen: primary particles
Primer: primers
Propanol: propanol
Prüfverfahren: test methods
Prüfvorschriften: test specifications
Pulverbeschichten: powder coating process
Pulverlacke: powder coatings
Pulverlackherstellung: powder coating production
Pulversprühen, elektrostatisch: powder spraying, electrostatic

PVK: pigment volume concentration
Pyknometer: pycnometer

Q

Qualitätskontrolle: quality control
Qualitätsmanagement: quality management
Quarzmehl: silica flour
Quellung: swelling
Q-Wert: q-value

R

Radikal: radical
Rakel: doctor blade
RAL-Register: RAL-register
Randwinkelmessung: contact angle measuring
Rauheitsbestimmung: roughness measurement
Raumtemperatur: room temperature
Rautiefe: roughness height
REACH (Registrierung, Evaluierung, Authorisierung von Chemikalien): Registration, Evaluation, Authorisation of Chemicals
Reaktion: reaction
Reaktionsgrund: catalyst primer
Reaktivverdünner: reactive thinners
Recycling: recycling
Reduktion: reduction
Reflektometer: reflectometer
Reflexionskurve: reflection curve
Refraktometer: refractometer
regenerieren: regeneration
Reibfestigkeit: scrub resistance
Reißlack: crackling finish
reizend: irritant
relative Luftfeuchtigkeit: relative humidity
Reproduzierbarkeit: reproducibility
Resit: resite
Resol: resole

Retention: retention
Rezeptierung: formulation development
Rezeptur: formulation
Rheologie: rheology
Rheologie-Additive: rheological additives
Richtrezepturen: suggested formulations
Ricinenöl: castor oil, dehydrated
Ricinolsäure: ricinolic acid
Ricinusöl: castor oil
Rieselfähigkeit: free-flowing properties
Ringleitungsstabilität: ring circuit stability
Rissbildung: cracking
Ritzhärte: scratch test
Rolle: roller
Rost: rust
Rostentfernung: rust removal
rostfreier Stahl: stainless steel
Rostschutz: rust protection
Rostschutzgrundierung: rust prevention primer
Rotationsviskosimeter: rotation viscometers
Rub-out-Test: rub out test
Rührer: stirrers
Rührwerksmühlen: attrition mills
Runzelbildung: wrinkling
Runzellacke: wrinkle finishes
Ruß: carbon black
Rutil: rutile
Rutscher: sanding machines, manual

S

Säurebeständigkeit: acid resistance
Säurehärtung: acid curing
Säuren: acids
Säurezahl: acid value
Salzausblühung: salt efflorescence
Salzsprühtest: salt spray test
Sandmühle: sand mill
Sandstrahlen: sand blasting

Sauerstoffinhibierung: inhibitory effect of oxygen
Schaum: foam
Schellack: shellac
Schergeschwindigkeit: shear rate
Scheuerfestigkeit: wet abrasion resistance
Schichtdickenmessung: film thickness measurement
Schlagtiefung: impact test
Schleierbildung: haze
Schleifbarkeit: sanding properties, grindability
schleifen: sanding
Schleifpapier: sanding paper
Schleifstaub: sanding dust
Schmelzbereich: melting range
Schmelzpunkt: melting point
Schmierpunkt: smear point
schmutzabweisend: dirt repellent
Schrumpfung: shrinkage
Schubspannung: stress
Schüttelsieb: vibrating screen
Schüttvolumen: bulking value
schuppenförming: plate-like
Schutzausrüstung: protective equipment
Schutzbrille: protective goggles
Schutzhandschuhe: protective gloves
Schwarzpigmente: black pigments
Schwermetalle: heavy metals
Schwerspat: baryte
Schwitzwassertest: saturated atmosphere test
Sedimentation: settling
Seeklima: maritime climate
SH-Lack: acid curing coating
Shop-Primer: shop primer
Sicherheitsdatenblatt: safety data sheet
Sicherheitsschuhe: protective shoes
sichtbares Licht: visible light
Sieb: sieve
Siebanalyse: sieve analysis
sieben: screening
Siedetemperatur: boiling temperature

Sikkativierung: driers, dosing of
Siliciumdioxid: silicone dioxide
Silicone: silicones
Siliconharze: silicone resins
Siliconöle: silicone oils
Siliconpolyester: silicone polyester resins
Slip-Additive: slip additives
Slurry-Verfahren: slurry process
Sojaöl: soya bean oil
Solvatation: solvation
Solvatationshülle: solvation wrapping
Sondermüllverbrennung: special waste incineration
Sonnenblumenöl: sun flower oil
Spachtel: putty
Spannungsreihe, elektrochemische: electrochemical series
Spektralfarben: spectrum colours
Spektralphotometer: spectrophotometer
Sperrgrund: sealer
Spezifikationen: specifications
spezifische Oberfläche: surface area, specific
Spinellpigmente: spinel pigments
Spreitung: spreading
Spritzapplikation: spray application
Spritzautomaten: spraying automatic devices
Spritzbild: spraying pattern
Spritzkabine: spray booth
Spritznebel: spray dust
Spritzpistolen: spray guns
Spritzstand: spraying station
Spritzverdünnung: spraying thinner
Spritzverfahren: spray application processes
Sprühdosenlacke: aerosol lacquers
Stabilisierung (von Dispersionen): stabilisation (of dispersions)
Stahl: steel
Stammlack: base material
Standardisierung: standardisation
Standardmuster: standard sample
Standfestigkeit: sagging resistance
Standöl: stand oil

Staubbindetuch: anti-static cloth
Staubexplosion: dust explosion
staubfrei: dust-free
Staubmaske: dust mask
Staubtrocknung: dust-free drying
Stearinsäure: stearic acid
Steinschlagschutz: stone chipping protection
Steinschlagtest: stone chipping test
Stippenbildung: speck formation
Strahlenhärtung: radiation curing
Strahlung: radiation
Straßenmarkierungsfarben: road line paints
Streichen: brushing
Streuvermögen: scattering power
Strukturlacke: structured paints
Strukturviskosität: pseudoplasticity
Styrol: styrene
Styrolalkydharze: styrenated alkyd resins
Substrat: substrate
subtraktive Farbmischung: subtractive colour mixture
Synthese: synthesis
synthetisch: synthetic

T

Tagesleuchtfarben: daylight fluorescent paints
Tagesleuchtpigmente: daylight fluorescent pigments
Talkum: talc
Tallöl: tall oil
Tanklager: tank farm
Tarnfarben: camouflage paints
Tauchlackieren: dipping
Taupunkt: dew point
Technikum: pilot plant
Teilchen: particle Teilchenform: particle shape
Teilchengröße: particle size
Temperatur: temperature
Tenside: surfactants
Terephthalsäure: terephthalic acid

Terpenkohlenwasserstoffe: terpenoid hydrocarbons
Terpentinersatz: turpentine oil substitute
Terpentinöl: turpentine oil
Testbenzin: white spirit
Thermoplaste: thermoplastic resins
Thixotropie: thixotropy
Thixotropierungsmittel: thixotropic agents
Titandioxid: titanium dioxide
Titration: titration
Toleranz: tolerance
Toluol: toluene
Toluolsulfonsäure: toluolsulphonic acid
Topfzeit: pot life
Totenkopf: skull and crossbones
Toxizität: toxicity
Tränklacke: impregnating varnishes
Träufelharze: dripping resins
Transparenz: transparency
Treibgas: propellant gas
Trennmittel: release agents
Tribo-Applikation: tribo-application
Trimellithsäure: trimellitic acid
Trimethylolpropan: trimethylol propane
Tristimulusverfahren: tristimulus method
Tri-Tauchverfahren: trilene dipping process
Trockenfilmdicke: dry film thickness
trockenschleifen: dry sanding
Trockenstoffe: driers
trocknende Öle: drying oils
Trocknung, chemisch: drying, chemical
Trocknung, forciert: drying, forced
Trocknung, oxidativ: drying, oxidative
Trocknung, physikalisch: drying, physical
Trocknungsprüfung: drying time measurement
Trocknungsund Härtungsverfahren: drying and hardening methods
Trommellackierverfahren: barrel coating
Trübung: haze
Tubenemaillen: tube enamels

U

überbrennen: overstoving
Überlackierbarkeit: overcoatability
Überstreichbarkeit: overcoatability
Übervernetzung: overcuring
Ultrafiltration: ultrafiltration
Ultramarin-Pigmente: ultramarine pigments
ultraviolette Strahlung: ultraviolet radiation
Umdrehungszahl: revolutions per minute
Umgriff: throwing power
Umluftofen: ciculating air oven
umweltfreundlich: eco-friendly
Umweltschutz: environmental protection
unbunte Farbe: achromatic colour
ungesättigte Fettsäuren: unsaturated fatty acids
ungesättigte Polyester: unsaturated polyesters
Uni-Lackierung: plain colour coating
Unterbodenschutz: underbody sealing
Untergrund: substrate
Untergrundvorbehandlung: pre-treatment of substrates
Unterrostung: sub-surface rusting
Untervernetzung: undercuring
Unterwasseranstriche: underwater coatings
Unverträglichkeit: incompatibility
Urethan: urethane
UV-Absorber: UV-absorbers
UV-Härtung: UV-curing

V

Vakuum-Dissolver: vacuum dissolver
Vakuum-Imprägnierung: vacuum impregnation
Van-der-Waalsche Kräfte: Van der Waals forces
Vapour-Cure-Verfahren: vapour cure process
Ventilationsgrund: ventilation surfacer
Verarbeitungsviskosität: application viscosity
Verbindung (chemische): compound (chemical)
Verbrennung: combustion

Verdicker: thickener
Verdünnbarkeit: thinnability
Verdünnungsmittel: thinner
verdunsten: evaporate
Verdunstungskälte: cooling by evaporation
Verdunstungszahl: evaporation index
Veresterung: esterification
Veretherung: etherification
Verfilmung: film-forming
Vergilbung: yellowing
verkappt, chemisch: blocked, chemically
Verkieselung: silicification
Verkochung: boiling
Verlacken von Farbstoffen: pigment lakes
Verlauf: levelling
Verlaufsmittel: levelling agents
Vernetzung: cross-linking
Vernetzungsdichte: cross-linking density
Vernetzungsgeschwindigkeit: cross-linking velocity
Vernetzungsreaktion: cross-linking reaction
Verschnittmittel: extenders
Verseifbarkeit: saponifiability
Verseifungszahl: saponification value
Versprödung: embrittlement
Verträglichkeit: compatibility
verzinken: zinc (to)
Vinylacetat: vinyl acetate
Vinylchlorid: vinyl chloride
Vinylpropionat: vinyl propionate
Viskosimeter: viscometers
Viskosität: viscosity
Viskositätsanomalien: viscosity anomalies
visuelle Beurteilung: visual evaluation
VOC-Richtlinie: VOC-directive
Volltonfarbe: full-tone colour, mass tone
Vorbehandlungsverfahren: pre-treatment methods
Vordispergieren: predispersion
Vorlack: undercoat varnish
Vossen Blau: Vossen blue

W

Wachse: waxes
Wärmehärtung: thermosetting
wässrige Dispersion: aqueous dispersion
Walzlackierung: roller coating
Wandfarben: wall paints
Wash-Primer: wash primer
Wasser, deionisiert: water, desalinated
wasserabweisend: water repellent (hydrophobic)
Wasseraufbereitung: water treatment
Wasserberg: water mountain
Wasserberieselung: wet rinsing systems
Wasserbeständigkeit: water resistance
Wasserdampfdurchlässigkeit: water vapour permeability
wasserfrei: water-free
Wasserglasanstrichfarben: water glass paints
Wasserhärte: water hardness
Wasserlacke: water-borne coatings
Wasserlöslichkeit: water solubility
Wasserstoffbrückenbindung: hydrogen bridge bond
Wasserstoffionenkonzentration: hydrogen ions concentration
Wasserverdünnbarkeit: water thinnability
Weather-O-Meter (WOM): weather-o-meter
Wechselklimatest: alternating atmosphere test
Weichharze: soft resins
Weichmacher: plasticisers
Weichmachung, äußere: plastification, external
Weichmachung, innere: plastification, internal
Weißabmischung: mixture with white
Weißanlaufen: blushing
weiße Ware: white goods
Weißpigmente: white pigments
Weißrost: white rust
Weißstandard: white standard
Wellenlänge: wavelength
Wetterbeständigkeit: weather resistance
Widerstand, elektrischer: resistance, electrical
Windsichten: air separation

Wirbelsinterverfahren: fluidised bed process
Wolkigkeit: clouding
Wollskala: wool scale
Wurzelharz: wood rosin

X

Xenontest: xenon test
Xylol: xylene
Xylol/Formaldeyd-Harze: xylene/formaldeyde resins

Z

Zaponlacke: zapon lacquers
Zementechtheit: cement resistance
Zeolithe: zeolites
Zersetzungstemperatur: decomposition temperature
Zerstäubung: atomisation
Zetapotential: zeta potential
Ziehen (von Spachteln): trowelling
Zinkchromat: zinc chromate
Zinkoxid: zinc oxide
Zinkphosphat: zinc phosphate
Zinkphosphatierung: zinc phosphating
Zinkstaub: zinc dust
Zinkstaubgrundierungen: zinc dust primers
Zinksulfid: zinc sulphide
Zinktetraoxichromat: zinc tetraoxychromate
Zinkweiß: zinc white
Zirkoniumtrockenstoff: zirconium drier
Zündtemperatur: ignition temperature
Zunder: scale
Zweikomponentensysteme: two-component, two-pack systems
Zweischichtlackierungen: two-coat systems
Zwischenschichthaftung: interlayer adhesion
Zwischenschliff: intermediate sanding

Author

After moving to Germany in 1966, **Paolo Nanetti** was appointed Coatings Technician in charge of the development of electrical insulation coatings at Herrmann Wiederhold Lackfabriken, Hilden/Germany, and then at Felten & Guillaume Dielektra, Porz/Germany. In 1971 he joined Spies & Hecker GmbH in Cologne/Germany, today Axalta Coatings Systems. First he worked on the development of putties and fillers (surfacers) and aerosol lacquers; later he became pigment advisor to the laboratories at the company's pigment centre, where he led the pigment basic research. In 1984 he took over the direction of the analytical laboratory, the testing of raw materials and the paste standardisation. In 1991 he assumed full time responsibility for the scientific apprenticeship and further education programs at the Cologne site. From 2002 to 2011 he worked as an Academic Research Assistant at the University of Applied Sciences Niederrhein, Krefeld/ Germany, in the Department of Coatings Engineering assisting students and executing projects. After retiring in 2011, he is still responsible for the education of laboratory assistants in the coatings industry.

www.ingramcontent.com/pod-product-compliance
Lightning Source LLC
Chambersburg PA
CBHW071223230926
51849CB00001B/16